AF847084

Springer Aerospace Technology

The *Springer Aerospace Technology* series is devoted to the technology of aircraft and spacecraft including design, construction, control and the science. The books present the fundamentals and applications in all fields related to aerospace engineering. The topics include aircraft, missiles, space vehicles, aircraft engines, propulsion units and related subjects.

More information about this series at http://www.springer.com/series/8613

Baburov S.V. · Bestugin A.R. ·
Galyamov A.M. · Sauta O.I. ·
Shatrakov Y.G.

Development of Navigation Technology for Flight Safety

Springer

Baburov S.V.
Saint-Petersburg, Russia

Galyamov A.M.
Moscow, Russia

Shatrakov Y.G.
Saint-Petersburg, Russia

Bestugin A.R.
Saint-Petersburg State University
of Aerospace Instrumentation (SUAI)
St. Petersburg, Russia

Sauta O.I.
Saint-Petersburg, Russia

ISSN 1869-1730 ISSN 1869-1749 (electronic)
Springer Aerospace Technology
ISBN 978-981-13-8374-8 ISBN 978-981-13-8375-5 (eBook)
https://doi.org/10.1007/978-981-13-8375-5

© Springer Nature Singapore Pte Ltd. 2020, corrected publication 2020
This work is subject to copyright. All rights are reserved by the Publisher, whether the whole or part of the material is concerned, specifically the rights of translation, reprinting, reuse of illustrations, recitation, broadcasting, reproduction on microfilms or in any other physical way, and transmission or information storage and retrieval, electronic adaptation, computer software, or by similar or dissimilar methodology now known or hereafter developed.
The use of general descriptive names, registered names, trademarks, service marks, etc. in this publication does not imply, even in the absence of a specific statement, that such names are exempt from the relevant protective laws and regulations and therefore free for general use.
The publisher, the authors and the editors are safe to assume that the advice and information in this book are believed to be true and accurate at the date of publication. Neither the publisher nor the authors or the editors give a warranty, expressed or implied, with respect to the material contained herein or for any errors or omissions that may have been made. The publisher remains neutral with regard to jurisdictional claims in published maps and institutional affiliations.

This Springer imprint is published by the registered company Springer Nature Singapore Pte Ltd.
The registered company address is: 152 Beach Road, #21-01/04 Gateway East, Singapore 189721, Singapore

The original version of the book was revised: The affiliation of the author "Bestugin A.R." has been updated. The correction to the book is available at https://doi.org/10.1007/978-981-13-8375-5_7

Introduction

The statistics of the world aviation accidents shows [1–6] that, despite significant efforts made by aviation equipment and avionics manufacturers to ensure flight safety (FS), a significant reduction in the number of accidents and incidents cannot be achieved. The material damage caused by aviation accidents around the world including those involving military aircraft (AC) amounts to tens of billions of dollars, and the non-pecuniary damage in civil aviation is extremely difficult to assess in general. That is why the issues of ensuring the AC flight safety are always the primary focus of the AC developers.

Guidance documents of the International Civil Aviation Organization (ICAO), the European Organisation for the Safety of Air Navigation (Eurocontrol), the Interstate Aviation Committee (IAC), the American Association of State Avionics Producers (RTCA), as well as the Operational Requirements (ORs) of various Russian forced to weapons means and systems determine the need to develop both special flight safety support systems and to improve the technical characteristics of the navigation systems used (including the GNSS) to improve the efficiency of AC use [7–9].

It is known [1–3] that the greatest number of accidents in aviation has occurred and still occurs at the landing phase during when aircraft collide with ground or artificial obstacles in reduced visibility conditions.

To improve the AC flight safety, in the 50s of the twentieth century, the development of meter-wave instrument landing systems has begun, now known as instrument landing systems (ILSs); and to prevent AC collisions with ground, hazardous descent warning systems appeared in the 1960s, such as Ground Proximity Warning System (GPWSs) in the USA and Hazardous Descent Warning Systems (HDWSs) in the USSR.

The development and implementation of instrument landing systems and collision avoidance systems made it possible to reduce the number of AC accidents; and, if not for the annual growth of the AC fleet and aerodrome network, the increase in the flying time and aging of the aviation fleet as a whole, the number of accidents and incidents would have been reduced by now to a large extent. Unfortunately, after a certain decrease in the total number of accidents resulting from the introduction of new systems, their growth has usually been observed, which is also true at present. Thus, a qualitative improvement in the situation with regard to reducing the accident rate still has not been achieved.

The need to improve the efficiency of the use of weapons and military equipment led to the development of the new generation of medium-altitude GNSSs (GLONASS and GPS) in the late 1970s by the military departments of the USSR and the USA, which provided ACs with accurate navigation information at any time in all regions of the world [10]. The high efficiency of GNSS applications in the military areas has also fueled the widespread introduction of GNSSs for civil applications since the 90s of the twentieth century. In the USA and Europe, Enhanced Ground Proximity Warning Systems (TAWS) [11] using GNSS information have been developed and introduced, and in the late 1990s, similar EGPWS systems appeared in Russia [12, 13]. At the beginning of the twenty-first century, instrumental landing systems appeared that used GNSS technologies: first in the USA, based on GPS [14, 15], and then in Russia, already using GLONASS and GPS [16–18].

The emergence of medium-altitude GNSSs (GLONASS, GPS) and the beginning of their practical use in military areas and in civil aviation resulted in a situation that ICAO is currently considering GNSS as the primary means of navigation at all phases of civil AC flights [18, 19].

The wide application of GNSSs is due to a number of both technical and economic factors. Their combination provides certain advantages of GNSSs in comparison with other radionavigation systems. Although GNSSs have some fundamental drawbacks, its wide practical use is explained by the fact that the elimination of these drawbacks for specific applications is possible in some inexpensive but effective manner.

A feature of the new systems for improving the AC flight safety, such as EGPWSs and GNSS-based landing systems (the international designation GLS—Global Landing System) are fundamentally new capabilities inherent in the GNSS system, which lead to a qualitative change in the approach to reducing the rate of aviation accidents and incidents.

Summarizing the above, one can conclude that one of the most important directions in the development of GNSS technologies is the integrated use of various systems that previously have not interacted, based on the GNSS principles.

The analysis of current trends in FS-related systems shows that despite the efforts being made, a contradiction has arisen and is growing today, which is the inconsistency between the constantly growing requirements for flight safety and the capabilities of existing radiotechnical systems for landing and collision avoidance. For example, the instrument radiobeacon landing system will not be able to prevent an AC from colliding with ground or an artificial obstacle at the aerodrome with a complex nature of the surrounding terrain. The limitations of the known approaches used in instrument landing systems and collision avoidance systems make it necessary to develop new methods, techniques and devices that allow both to increase the efficiency of each of these systems and to gain additional advantages in the development of GNSS technologies through the integrated use of these systems.

Thus, it is topical to develop and scientifically substantiate new methodological approaches and technical solutions that enable compliance with modern and future requirements to AC flight safety.

The results presented in the monograph made it possible to develop a new methodological approach to solving the problems of increasing the AC flight safety both by using GNSS technologies separately in collision avoidance systems (CASs) and in satellite-based landing systems (SLSs), and by applying new methods in integrated use of CASs and SLSs.

The use of the proposed approaches makes it possible to create and develop a new class of integrated systems, which provide for a substantial increase in military and civil aviation AC FS and also increase the economic efficiency of civil aviation.

The monograph addresses radiotechnical systems for providing AC FS, which use satellite technologies based on medium-altitude GNSSs, as well as methodological approaches and GNSS-based hardware–software complexes providing practical implementation of the proposed approaches to FS increasing.

The purpose of this monograph is to systematically present new scientifically substantiated methodological approaches and GNSS-based technical solutions that allow reducing the probability of accidents during military and civil AC flights.

The problem under consideration is of a complex nature and encompasses a wide range of scientific and technical directions. This includes a comprehensive analysis of the state and identification of the main problems in the development and operation of instrument landing systems and CASs, approaches to the construction of GNSS-based radiotechnical complexes, synthesis of structures for building SLSs and CASs, and recommendations for the creation of multifunctional systems.

The monograph presents new methods for increasing the accuracy, integrity, continuity, ergonomics, and availability of satellite-based landing systems and collision avoidance systems, in which GNSS technologies are used, including the conditions of electromagnetic interference. It also describes new methodical approaches to the design of SLS and CAS structures and new methods to improve flight safety based on the integrated use of these systems. The peculiarity of all the developed methods and structures for constructing radiotechnical complexes is the their applicability in other areas of science and technology related to highly accurate determination of the state parameters for the controlled object (road, sea, and river transport, military navigation systems of all branches of Armed Forces). Practically all the developed methods can be used where GNSS technologies are used and where it is necessary to provide high requirements to the accuracy, integrity, and continuity of navigation information.

The practical value of the monograph is that, based on the approaches described in it, it is possible to systematize and simplify the procedure for making decisions and to reduce the influence of subjective factors in selecting the main functional elements of navigation systems, including those using expert navigation-oriented systems with a wide range of preference criteria for multiple factor analysis. The use of the approaches proposed in the monograph makes it possible to significantly shorten the time frame for the development of new flight safety systems that use the existing structures of onboard navigation and flight systems on various AC types to the fullest extent.

Chapter 1 analyzes the existing landing systems and collision avoidance systems, their performances, and trends of their development. It briefly describes the GNSS structure and functions and specifies the reasons for the use of augmentations in satellite-based landing systems and collision avoidance systems. It also presents requirements for the performance of satellite-based landing systems and collision avoidance systems. Based on the analysis, prospective methods and means to improve flight safety in satellite-based landing systems and collision avoidance systems are considered. At the end of the chapter, safety performance indicators are selected, and the problem under consideration is formalized.

Chapter 2 presents the methodology for constructing satellite-based landing systems and collision avoidance systems; proposes an approach to selecting the basic elements of the structures of these systems; considers the structure of the satellite-based landing system based on GNSS augmentations, and the structure of the collision avoidance system using GNSS technologies. At the end of the chapter, directions and ways are presented that may be used to enhance the considered systems and to improve flight safety.

Chapter 3 presents methods to improve flight safety when using satellite-based landing systems. It also considers the methods of constructing and using diagrams of the volumetric distribution of radiowaves multipath propagation errors; the method of ensuring the integrity and continuity of information during AC approach and landing using the integrated signal-to-noise ratio for the measured pseudoranges; the method for compensating pseudorange errors in the SLS ground and onboard subsystems based on phase measurements; and the method of using pseudosatellites in satellite-based landing systems.

Chapter 4 presents methods for improving flight safety for collision avoidance systems. It considers the method of collision avoidance based on the three-dimensional synthesis of the cross sections of the underlying surface (ground) and the mapping of hazardous elements; the method of estimating the possibility of vertical maneuvering and determining the direction of the turn; the method for determining the hazardous terrain relief, taking into account the possibility of AC reverse; and the method of analyzing the space inside the corridor that is safe for the flight.

Chapter 5 presents comprehensive technical solutions for the joint use of satellite-based landing systems and collision avoidance systems based on GNSS technologies. It defines the principles of building an integrated flight safety system on the basis of satellite-based landing systems and collision avoidance systems. It proposes a method for preventing AC landing on an unauthorized runway by calculating a virtual glide path, and a method of reporting the AC position during landing and landing roll. At the end of the chapter, the improvement of flight safety with the use of integrated systems is assessed.

Chapter 6 provides recommendations on the application of the technical solutions proposed in the paper in satellite-based landing systems and collision avoidance systems and the principles of construction, and design features of onboard equipment for improving the AC flight safety. Examples are given of the construction of an onboard navigation and landing complex on the base of the satellite-based landing system and the collision avoidance system. The results of flight tests and practical operation are presented. At the end of the chapter, an integral assessment of flight safety improvement is made, and recommendations are given on the practical application of the proposed technical solutions.

In conclusion, the main results are provided that were achieved in the process of practical implementation of the methodological approaches, structures for the construction of systems, methods, and devices to improve flight safety that are presented in the monograph.

References

1. Annual report of the Council (2011) [Electronic resource]. International Civil Aviation Organization. Doc 9975. www.icao.int
2. Aralov GD (2012) Analysis of flight safety issues. In: Flight safety issues, No. 5, 2012. VNIITI, Moscow (in Russian)
3. Annual report of the Interstate Aviation Committee (IAC) "Flight Safety Status in 2011" [Electronic resource], www.mak.ru (in Russian)
4. Materials of the 5-th international conference "Safety of the air transport system". [Electronic resource]. Moscow, 20 Feb 2012 (in Russian)
5. Guziy AG, Lushkin AM (2008) Quantitative estimation of the current level of aircraft operator flight safety. In: Flight safety issues, No. 10, 2008. VNIITI, Moscow (in Russian)
6. Abstract of a study conducted by the European Aviation Safety Agency (EASA) for 2009 (2010) In: Flight safety issues, No. 11, 2010. VNIITI, Moscow (in Russian)
7. Military doctrine of the Russian Federation for the period until 2020/Approved by the Decree of the President of the Russian Federation of 5 Feb 2010, No. 146 "On the Military Doctrine of the Russian Federation". [Electronic resource]. http://stat.doc.mil.ru/documents (in Russian)
8. Safety management manual (2006) Doc 9859, First edition [Electronic resource]. International Civil Aviation Organization (in Russian)
9. Baburov VI, Rogova AA, Sobolev SP (2005) Method for calculating deviations from the heading line and glide path in the onboard equipment of the satellite-based landing system. NSTU Sci Bull 1:3–10 (in Russian)
10. Dmitriev PP et al (1993) Network satellite radio navigation systems. In: Shebshayevich VS (resp ed) 2nd edn. Radio and Communication, Moscow, 408 p (in Russian)
11. Terrain Awareness and Warning System (TAWS). ARINC 763, Recommendations dated 10 Dec 1999 (in Russian)
12. Certificate of the product conformance (2003) SGKI-034-112-SRPBZ. Interstate Aviation Committee. Aviation register (in Russian)
13. Dryagin DM (2006) Complex enhanced ground proximity warning system with enhanced functionality and software algorithms that minimize the false alarm probability [Electronic resource]. Thesis and Tech Sciences: 05.11.03. RSL, Moscow (in Russian)
14. Akos D (2000) Development and testing of the Stanford LAAS ground facility prototype. In: Akos D, Gleason S, Enge P, Luo M, Pervan B, Pullen S, Xie G, Yang J, Zhang J (eds) NTM 2000: Proceedings of 2000 national technical meeting of the institute of navigation. Anaheim, pp 210–219
15. Braff R (2001) LAAS Performance for terminal area navigation. In: ION 57th Annual meeting/CIGTF 20th Biennial guidance test symposium, Albuquerque, NM, pp 252–262, 11–13 June 2001

16. Solovyev YuA (2003) Satellite navigation and its applications. Eco-Trends, Moscow, 326 p (in Russian)
17. Yatsenkov VS (2005) Fundamentals of satellite navigation. NAVSTAR and GLONASS GPS systems. Goryachaya liniya-Telecom, Moscow, p 272 (in Russian)
18. Local Area Augmentation System. Certificate of type No. 399 of 02.12.2005. Interstate Aviation Committee. Aviation register (in Russian)
19. Annex 10 to the Convention on International Civil Aviation. Aeronautical Telecommunications. Radio Navigation Aids, vol 1, 6th edn, July 2006 (in Russian)

Contents

Abbreviations

ABAS	Aircraft-based augmentation system
AC	Aircraft
ADB	Aircraft database
ADF	Automatic direction finder
ADS	Air data system
ADS-B	Automatic dependent surveillance–broadcast
AFS	Antenna-feed system
AMS	Airborne multifunctional system
CAS	Collision avoidance system
DC	Differential corrections
DDRE	Differential data reception and conversion equipment
DH	Decision height
DL	Data link
EGPWS	Enhanced Ground Proximity Warning System
EWS	Emergency warning system
FAS	Final approach segment
FI	Flight intercom
FM	Frequency modulated
FS	Flight safety
GBAS	GNSS ground-based augmentation system
GISR	Ground integrated signal-to-noise ratio
GLONASS	GLObal NAvigation Satellite System (Russia)
GLS	GBAS landing system
GNSS	Global Navigation Satellite System (currently including GLONASS and GPS)
GNSS/LAAS	Onboard satellite-based landing equipment
GPS	Global Positioning System (USA)
GSR	Ground signal-to-noise ratio
HSI	Hardware/software interface
IAC	Interstate Aviation Committee

ICAO	International Civil Aviation Organization
ILS	Instrument landing system
LAAS	Local monitoring and correcting station
LAL	Lateral alert limit
LOC	Localizer
LPL	Lateral protection level
LRBG	Landing radar beacon group
MAA	Minimum allowable altitude
MDE	Minimum detectable error
MMR	Multi-mode receiver
MP	Multipath
MPD	Multipurpose display
NPA	"Non-precision" approach
NSV	Navigation space vehicle
OISR	Onboard integrated signal-to-noise ratio
PLL	Phase-lock loop
PPRNS	Pulse-phase radionavigation systems
PR	Pseudorange
PRNS	Phase radionavigation systems
PS	Pseudosattelite
RDB	Earth relief database
REC	Radio-electronic complex
RR	GNSS reference receiver
RWY	Runway
SBAS	GNSS satellite-based augmentation system
SLS	Satellite-based landing system
SV	Space vehicle
UAV	Unmanned aerial vehicle
UTC	Coordinated Universal Time
VAL	Vertical alert limit
VDB	VHF band digital signal for data transmission
VOR	VHF omnidirectional radiorange
VPL	Vertical protection level

Chapter 1
General Description of Flight Safety Problems

One of the most important problems that military and civil aviation faces is to ensure landings with a target level of safety in any weather conditions in any region of the globe. The solution to this problem depends primarily on the AC and aerodrome equipment with instrument landing systems [1–4].

At the same time, ensuring flight safety is a complex task, the solution of which requires the use of other means, such as collision avoidance systems [5–12].

The development and commissioning of these systems provided a significant reduction in the accident rate. However, to ensure compliance with modern safety requirements, further improvement is required.

1.1 Analysis of the State and Prospects for the Development of Instrument Landing Systems and Collision Avoidance Systems

1.1.1 Instrument Landing Systems

Currently, the most widespread of all instrument landing systems are radio technical systems based on the use of ground-based radio beacons [2, 4, 13], which form some fixed distribution of the electromagnetic field (the directional pattern) in space, on the basis of which it is possible to determine AC deviations from the heading plane and glide paths during the approach phase. These systems belong to the class of meter-wave radio beacon systems and have the international designation “ILS” (Instrument Landing System) [13].

Other types of radio technical landing systems, such as radar (RBS) and those with scanning beam (microwave landing system), are not so widespread, mainly because of their high cost.

The characteristics of various types of radio technical landing systems (LSs) are detailed in the references [1–4, 13].

© Springer Nature Singapore Pte Ltd. 2020

Baburov S.V. et al., *Development of Navigation Technology for Flight Safety*, Springer Aerospace Technology, https://doi.org/10.1007/978-981-13-8375-5_1

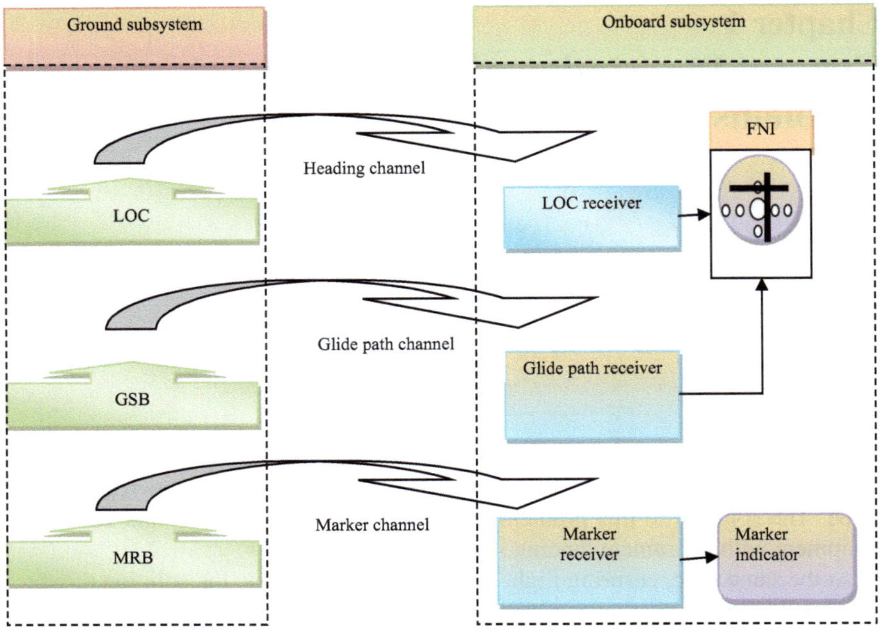

Fig. 1.1 Instrument landing system structure

The main advantage of systems such as ILS is their wide use, sufficiently high accuracy of the formation of guidance signals, and reliability. The drawbacks of these systems include the high cost of equipment and operation.

A general landing system of the ILS type consists of two radio beacons: localizer (LOC) and glide slope beacon (GSB) [1].

The LS generalized schematic diagram is shown in Fig. 1.1.

The range of LOC operating frequencies is 108–112 MHz. The LOC antenna system forms simultaneously two horizontal radiation diagrams in the space. As a result of the summation of signals, the total field is distributed in space in such a way that there is no error signal along the runway centerline, and in case of a deviation from the centerline, the signal increases in the positive or negative direction.

The LS onboard equipment measuring the magnitude of the corresponding signal determines the side and angle of AC deviation from the plane passing through the runway centerline.

The range of GSB operating frequencies is 329–335 MHz. The directional pattern of the GSB antenna system is formed as a result of reflection of radio waves from the ground surface; therefore, there are increased requirements to the underlying surface directly adjacent to the GSB antenna system. The GSB uses the same working principle as the LOC.

The intersection of the heading plane and the glide slope plane gives a glide slope line. Due to uneven terrain and obstacles in the coverage area of the radio beacons, the glide slope line is subject to curvatures, the magnitude of which is normalized for each category of the landing system. The glide path angle according to ICAO requirements is chosen in the range of 2°–4° [1].

For ease of use and monitoring the approach at aerodromes, marker beacons (MRB) are used that are located on the extended runway centerline from the approach side. In Russian LSs, two MRBs are usually used (long-range—LRB and short-range—SRB). The LRB and SRB radiate the radio signal upward in a narrow spatial sector. Their signals will be received onboard the aircraft only if the aircraft does not deviate significantly from the heading plane. When the aircraft flies over MRB, an alert system is activated. The alert of the SRB overflight informs the pilot that he/she is in the immediate vicinity of the runway and must decide whether to land or execute a go-around.

In fact, LSs such as ILS form two virtual planes in the space in front of the runway (RWY): heading and glide slope. The intersection of these planes defines a line in the space called the glide slope line. The pilot's task is to keep the aircraft on this line while performing the approach. Thus, LSs are, in fact, systems providing guidance to the runway.

At the end of the 50s of the twentieth century, the USSR developed a short-range radio technical navigation system (SRNS), which included a landing radio beacon group (LRBG) and a corresponding onboard subsystem [1, 3, 14–17]. In this landing system operating in the 800–900 MHz band with the same principles as ILS, the effect of multipath propagation from the earth's surface has been significantly reduced. However, the system was used only at military airfields and on military aircraft, and its cost was still high.

In order to overcome the main drawbacks of the meter-wave LSs, in the 1970s, the USA, Europe, and the USSR began the development and implementation of a new landing system that operated in the centimeter range of waves. The system was named MLS—microwave landing system [1–3]. The MLS system does not experience problems with reflections of radio signals from the earth's surface, airfield structures, and other aircraft in the areas of antenna radiation. It determines the AC position more accurately at a wide range of angles in front of the runway. When using the MLS, it becomes possible to perform approaches along curved trajectories, safety intervals are reduced, and the airport capacity in adverse weather conditions is increased. However, the cost of the ground and onboard parts of the MLS could not be significantly reduced in comparison with the ILS equipment, and this fact prevented airlines and airports from investments in the implementation of this system.

Radar landing systems (RLSs) [1, 3] were not widely used in civil aviation to provide instrument approaches, also because of the high cost of the ground and onboard equipment. RLSs are currently mainly used in air traffic control (ATC) centers to monitor approaches.

The complexity and high cost of equipping aerodromes and aircraft with instrument landing systems still hamper the reduction in the number of accidents and incidents due to the fact that, for example, in Russia, only 100 out of 1500 operating aerodromes are equipped with instrument landing systems.

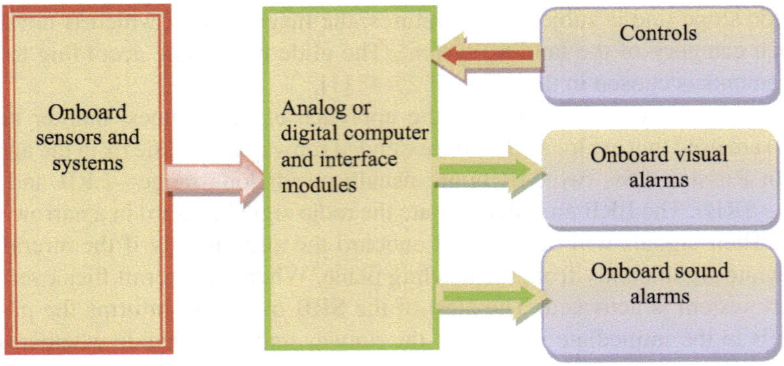

Fig. 1.2 Collision avoidance system structure

For the normal LS operation, airports impose constraints on the movement of aircraft on the ground so that they do not obscure and do not reflect the LOC and GSB signals, and this reduces the capacity of airports, especially when operating under adverse weather conditions. Traditional LSs can serve only for direct approaches from one side of the runway, since there is only one line of equal intensity of the beacons. At the same time, in many airports the nature of the terrain requires a more complex approach, which can be performed with MLSs.

Conventional ILS and MLS approach technologies are currently being developed mainly by upgrading the onboard equipment, including its integration with the GNSS equipment [17–19].

Simultaneously with the development of instrument landing systems, collision avoidance systems were created and improved.

Collision avoidance systems. The first systems of this class were collision avoidance systems, which in ICAO terminology are referred to as ground proximity warning system (GPWS) developed in the USA [8] and in the USSR [20–22].

The generalized GPWS diagram is shown in Fig. 1.2. The most important GPWS element is the computer. In it, based on the current values of the signals from the onboard radio altimeter, the pressure altitude sensor, the air speed sensor, the LS onboard radio receiver, and also depending on the position of the gear and flaps, the necessary functions (modes) are implemented and hazardous proximity warnings are generated. These signals are issued to the onboard visual and sound alarms located in the cockpit.

According to the statistics of the 1970s, about eight commercial jet planes crashed every year because of a collision with the underlying surface (according to international terminology, such accidents are called CFIT—controlled flight into terrain) [23, 24], and these accidents occurred with fully controlled aircraft piloted by a highly qualified flight crew.

Accidents classified as CFIT are the main source of flight accidents [25]; during the period from 1988 to 1995, 2,200 people died in 37 accidents [26, 27].

Investigations have shown that the main causes of the tragedies were bad weather conditions, navigation errors, hazardous (challenging) terrain, and communication problems. For 30 years, in the process of improving the element base and the development of computer technology, systems have been improved, but the set of basic functions of these systems has remained unchanged.

Despite the fact that the introduction of GPWS systems significantly reduced the number of CFIT accidents, they were not completely eliminated (approximately 35% of all CFIT accidents occurred with aircraft with an installed and functioning warning system). The main reasons for these accidents were as follows: absence of alarms (28%); late output of alarms and insufficient time for the pilot to correct the situation (36%); and inadequate and delayed actions of the flight crew (40%) [28].

Significant progress in reducing aviation accidents has been achieved after the introduction of new functions that have eliminated the main drawbacks of GPWS class systems, such as the delayed output of alarms and lack of AC protection from collisions with the underlying surface and artificial obstacles. These new functions include the functions of "forward terrain assessment", "premature descent warning", and "showing of the underlying surface nature on the display".

This additional functionality is based on the use of digital databases of the underlying surface, artificial obstacles and the aerodrome database, and, most importantly, the use of information from the GNSS [29].

The emergence and widespread introduction of the GNSS in aviation has created a real alternative to conventional radio navigation facilities and has greatly accelerated the development of ground proximity warning systems, in which the early warning function has been implemented including the new functions listed above. The emergence and widespread development of the GNSS in present makes it possible to suggest new solutions to meet the safety requirements in the terminal area and during AC landings, and at the same time to transit to the use of relatively inexpensive equipment. The latter is extremely important for equipping both numerous poorly equipped aerodromes and landing areas, and the entire AC fleet [30].

The use of the GNSS in military aviation for high-speed aircraft [31] opens up new opportunities in terms of increasing the mobility and effectiveness of the use of aircraft, given that currently economic factors play an increasingly important role in the implementation of military doctrines of the Russian Federation [32, 33].

The use of the GNSS makes it possible to integrate such previously independent and non-interacting systems as satellite-based landing systems (SLSs) and collision avoidance systems (CASs) and to ensure their functioning on the basis of common principles based on GNSS technologies.

In this regard, it is expedient to briefly review the main functions, features of the construction, and use of the GNSS.

1.2 Features of Global Navigation Satellite Systems as an Instrument Basis for Improving Flight Safety

A simplified diagram of the GNSS construction is shown in Fig. 1.3 and includes: a space port, a space vehicle (SV) system, user's equipment, a command and measurement complex (CMC), and a control center [34–37]. The space port provides placing SV into the required orbits during the GNSS deployment, as well as periodic replenishment of the SV number as they reach the end of their life span.

The SV system is a set of navigation satellites (SVs) transmitting information to users. The satellites have means of spatial stabilization, equipment for trajectory measurements, telemetry, command and control, as well as power supply and thermal control systems.

The command and measurement complex (CMC) serves to supply the SVs with the service information required for navigation sessions, monitoring, and control.

The users' equipment is designed to receive and process SV signals; for this purpose, a dedicated computing device is provided in the receiver that solves the navigation task.

The GNSS operation is based on measuring the distance between the phase centers of the user equipment antenna and the antennas of the space vehicles, the position of which is known with high accuracy [34, 36]. The method of measuring the distance from the SV to the receiver antenna is based on the constancy of the propagation velocity of radio waves in space. In addition to ephemeris, the satellite transmits time stamps that allow unambiguous estimation of the receiver timescale offset relative to the GNSS system time.

Let us introduce the following notation: R_i^{meas} is the measured range from the receiver to the ith navigation satellite; Δt_i is the propagation time of the signal on

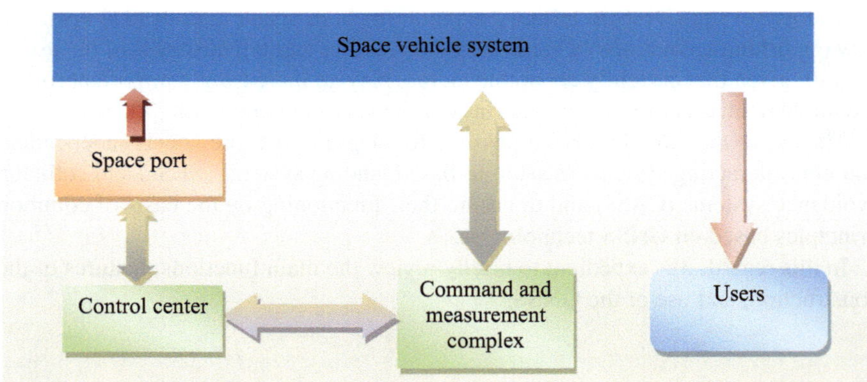

Fig. 1.3 Generalized GNSS diagram

the "ith satellite-user" route at the time of the navigation measurement; and c is the velocity of propagation of electromagnetic waves in space.

Then, the distance from the user to the ith SV is defined as:

$$R_i^{\text{meas}} = c\Delta t_i. \qquad (1.1)$$

Equation (1.1) can also be written in terms of the coordinates of the ith SV (x_i, y_i, z_i) and the user's coordinates (x, y, z) to be measured in the Cartesian system:

$$R_i^{\text{meas}} = \sqrt{(x - x_i)^2 + (y - y_i)^2 + (z - z_i)^2}, \quad i = 1,\ 2,\ \ldots \qquad (1.2)$$

To solve Eq. (1.2), it is necessary to measure ranges of up to three satellites, after which a system of three nonlinear equations with three unknowns is solved.

However, due to the fact that the SV and the user timescales are not initially synchronized, an error appears in determining the ranges from Eq. (1.2) because of their discrepancy, but since all ranges are measured instantaneously and the timescales of the navigation satellites are synchronized, the discrepancy of the "SV-user" scales at the moment of determining the ranges can be considered as a constant unknown value to be estimated along with the user's coordinates. Let us set this unknown value $= h_\tau$. Then, the system of Eq. (1.2) can be written as follows:

$$R_i^{\text{meas}} = \sqrt{(x - x_i)^2 + (y - y_i)^2 + (z - z_i)^2} + c \cdot h_\tau. \qquad (1.3)$$

Equation (1.3) contains four unknown values x, y, z, h_τ, and to solve it, ranges of at least four satellites are required.

The result of the solution of the system (1.3) with $i = 1, 2, 3 \ldots$ is the user's coordinates x, y, z and the discrepancy of the SV network and the user's equipment timescales h_τ [34, 36].

The discrepancy between the timescales of the navigation satellite network and the user clock is one of the sources of errors in determining the ranges.

In a more general form, the system of Eq. (1.3) can be written as follows:

$$R_i^{\text{meas}} = \sqrt{(x - x_i)^2 + (y - y_i)^2 + (z - z_i)^2} + c \cdot h_\tau + \Delta_i, \quad i = 1,\ 2,\ \ldots \qquad (1.4)$$

where Δ_i are errors in determining the range to the ith SV due to inaccuracy in the prediction of the ephemeris, errors in the velocity of propagation of radio waves in the troposphere and ionosphere on the "ith SV-user" routes, multipath errors of the navigation SV signals at the receiving site, noise in the user equipment reception channel and natural or intentional interference.

Based on the GNSS general structure and functions, let us briefly consider the features of building the existing varieties of Russian and foreign global navigation satellite systems.

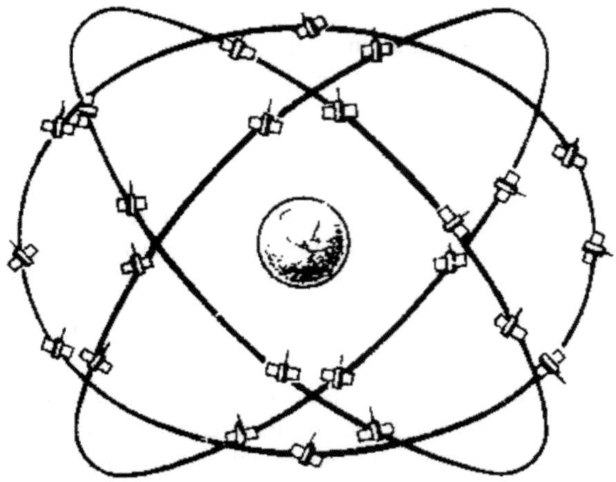

Fig. 1.4 GLONASS space vehicle system

GLONASS. The structure and functions of the Russian GLONASS system generally correspond to the general GNSS structure (see Fig. 1.3). Features include technical characteristics and the principle of the formation of navigation signals. The orbital constellation includes 24 SVs placed in three orbits offset along the equator by 120°, eight satellites in each, and is characterized by the following parameters: circular orbit, altitude of 19,100 km, revolution period of 11 h 15 min, and orbit plane inclination of 64.8° (Fig. 1.4) [34, 37, 38].

The method for dividing the signals emitted by the GLONASS SVs is a frequency one. The SV signals are identified by the value of the carrier frequency nominal lying in the allocated frequency band. There are two frequency bands in the L1 (~1.6 GHz) and L2 (~1.24 GHz) bands provided. Frequency nominals are formed according to the general rule:

$$f_{ij} = f_j + i\ F_{0j}\ , \tag{1.5}$$

where f_{ij} are the letter frequency nominals, f_j is the first letter frequency, F_{0j} is the interval between letter frequencies, $i = 0, 1, \ldots 24$ are letter numbers in each of the bands.

For civil users of the GLONASS system, all SVs emit radio signals modulated with a ranging code and service information in the L1 and L2 bands. At the same time, radio signals intended for military use are transmitted, which are modulated by a special code.

The equipment of GLONASS system users in the navigation session makes measurements of range and radial velocity of at least four SVs without requests. Based on the measurements of the radio navigation parameters and the service information extracted from the frame, the spatial coordinates (position) of the user are determined that make up the speed of movement, and the correction of the local timescale to the GLONASS timescale [37].

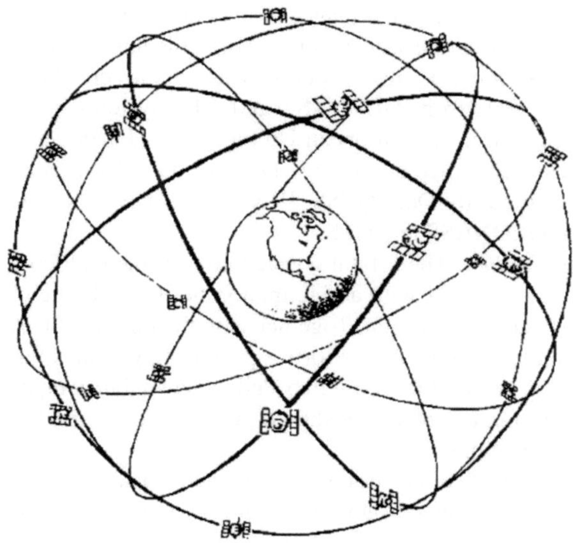

Fig. 1.5 GPS space vehicle system

GPS. The structure of the US GPS system is also constructed in accordance with the general GNSS structure (see Fig. 1.3). The space vehicle system includes a minimum of 24 satellites, ensuring a global high-accuracy navigation definition at any point of time.

The GPS configuration is a collection of six circular orbits at an altitude of about 20,000 km, four satellites in each orbit. Figure 1.5 shows a general view of the space GPS constellation of satellites [34]. Each GPS satellite emits signals modulated with two codes: the public C/A, which is intended for civil users, and the protected P used only by authorized users. Both codes are transmitted at a common frequency $f_1 = 1575.42$ MHz, with two carrier components offset by $\pi/2$ to eliminate amplitude modulation. The P and C/A codes are time synchronized and are distance-measuring, i.e., serve to measure pseudoranges.

To transmit the service information, the code D (Data) is used, which modulates the two carriers [34, 39].

The GPS system uses a code-based method for dividing signals from various SVs. The codes are formed by two pseudorandom sequence (PRS) generators; the selection of the initial shift register of one of the codes distinguishes the generated code sequence of this *i*th satellite. Of the large number of possible states, only 32 are selected, which generate codes with the best characteristics in the time-frequency plane [34]. Thus, it becomes possible to identify by code all satellites of the system, the number of which, not counting the backup ones, is 24.

The user equipment measures the SV pseudorange by estimating the delay of the range PRS and the radial velocity by estimating the Doppler offset of the carrier frequency. P and C/A signals are encoded with appropriate service information containing ephemeris, almanac, frequency-time corrections, time stamps, and information on the performance of onboard equipment. With these measurement results, when using the service information, the navigation-time task is solved.

At present, many countries are deploying their own GNSSs to ensure independence from GLONASS and GPS systems controlled by the relevant defense ministries. The European Union, China, Japan, and India are conducting intensive research aimed at creating their own GNSSs. It is assumed that these new systems (Galileo, BeiDou, QZSS, IRNSS) will have characteristics corresponding to those of GLONASS and GPS.

Galileo. The orbital constellation of the Galileo system will be located in three circular orbits with an inclination of 56° and a height of 23,616 km, 9 SVs plus a backup one on each. Together with the upgraded GPS and GLONASS systems, the Galileo system should form a promising global navigation satellite system. Galileo, GPS, and GLONASS will be independent, but compatible and interoperable systems, the joint use of which should provide the required service characteristics for many applications.

In the Galileo system, it is planned to use 10 navigation signals with right-handed circular polarization in the frequency ranges of 1164–1215 MHz, 1215–1300 MHz, and 1559–1592 MHz and one signal for providing search and rescue functions in cooperation with the Cospas-Sarsat system.

All SVs of the Galileo system will operate at the same frequencies; and to distinguish the SV signals, the principle of code division of channels, CDMA, will be used.

COMPASS. The PRC's own navigation system for the countries of Southeast Asia and the Pacific is being deployed and by 2015 will be transformed into a full-fledged GNSS with a system of 25 SVs. The system will consist of four geostationary satellites (GEOs), 12 SVs on inclined geosynchronous orbits, and nine SVs on circular orbits with a height of 22,000 km. The distance-measuring codes of the COMPASS system are identical in structure to the signals of the GPS system and contain a component with and without data.

Currently, the global navigation satellite system, which includes two systems GLONASS (Russia) and GPS (USA), is an international standard navigation aid [13] used by aviation users of various types and departments. This means that, firstly, the GNSS must meet the requirements specified in the international standards and, secondly, that any changes or additions to the relevant standards can be made on the basis of prior notification, at least six years before they are implemented.

The position information provided to all GNSS users, including aviation users, is expressed in geodetic coordinates of the World Geodetic System—1984 (WGS-84). Time data is expressed in the Coordinated Universal Time (UTC) system.

Let us briefly dwell on the requirements that are imposed on the basic functions and technical characteristics of the aircraft onboard navigation and information systems using GNSS data.

In the modern navigation concept developed by ICAO and based on performance (PBN) [40], it is stated that the requirements for the performance of the onboard area navigation system (RNAV) shall be defined in terms of accuracy, integrity, availability, continuity, and functionality, necessary for the performance of the intended flights in the context of the concept of a specific airspace. The PBN concept is a transition from navigation based on the use of specific sensors to performance-based navigation. The performance requirements are specified in the navigation specifications, which also determine which navigation sensors and equipment can be used to meet the requirements for the specified performance. These navigation specifications are described in sufficient detail to ensure coherence at the global level by providing states and aircraft operators with guidance on the implementation of specific equipment, including onboard equipment.

The characteristics of the GNSS signal in space are described in detail in [13] under the assumption of using the concept of a "fail-safe" user receiver and without taking into account such sources of errors as ionosphere, troposphere, interference, receiver noise, and multipath. This theoretical receiver is used only as a means of determining the performance for combinations of various GNSS elements. It is assumed that the "fail-safe" receiver shall be a receiver with nominal performance of accuracy and time-to-alert, and that such a receiver shall not have failures that affect integrity, availability, and continuity.

Let us consider the basic requirements from aviation users to the performance of the onboard equipment and compare them with the performance potentially provided by the GNSS system.

Accuracy. Requirements for the accuracy of the user equipment using standard accuracy signals from navigation GLONASS SVs are specified in regulatory documents [13, 38, 40]. It is assumed that to determine navigation parameters in the onboard GNSS equipment, SVs with a mask angle of 5 are used, and the user is in the near-Earth space that extends up to an altitude of 2000 km above the Earth's surface.

The error in determining the horizontal coordinates (positioning) in the onboard satellite navigation equipment (OSNE) [41] shall be no worse than 32 m (with probability 0.95) when the aircraft moves with a horizontal acceleration of not more than 5.7 m/s^2 and the acceleration rate of not more than 2.5 m/c^3. At the same time, the geometric factor HDOP, which characterizes the SV coordinates in the sky used to determine the position, shall be no more than 1.5.

For the "fail-safe" user receiver, [13] states that position errors in the horizontal plane of the GLONASS standard accuracy channel do not exceed 19 m (the global average for 95% of the time) and do not exceed 44 m for the worst site in the near-Earth space (for 95% time).

The error in determining the altitude above the surface of the reference ellipsoid in the OSNE [13] shall be no worse than 66 m (with probability 0.95) when the aircraft moves with a vertical acceleration of not more than 4.9 m/s^2 and the acceleration rate of not more than 2.5 m/c^3. In this case, the geometric factor VDOP, which characterizes the NSV coordinates in the sky used to determine the position, shall be no more than 3.0.

For the "fail-safe" user receiver, [13] states that position errors in the vertical plane of the GLONASS standard accuracy channel do not exceed 29 m (the global average for 95% of the time) and do not exceed 93 m for the worst site in the near-Earth space (for 95% time).

In the OSNE, the ground speed with a probability of 0.95 in the flight conditions specified above shall be determined with an error of not more than 0.3 m/s, and the track angle with a probability of 0.95 with a change in the ground speed from 100 to 1200 km/h in the flight conditions specified above shall be determined with an error of 35 to 3 angular minutes.

The current UTC time shall be output by the OSNE with an error of not more than 0.001 s with a probability of 0.95. For the "fail-safe" user receiver, it was determined in [13] that errors in the transmission of time data in the GLONASS standard accuracy channel do not exceed 700 ns for 95% of the time.

Continuity. Continuity of the service for the system is the capability of the system to perform its functions during the intended operation without unscheduled interruptions. Continuity of the service is the average probability that during a given period, the characteristics of the parameters are within the established tolerances [13].

The OSNE shall provide the continuity of data output at all stages of the aircraft flight, except for precise approaches, with a value not worse than 1×10^{-4}/h, and for precise approaches with a value not worse than $1–8 \times 10^{-8}$ for 15 s.

For the "fail-safe" user receiver, it is determined in [13] that continuity shall be no worse than 1×10^{-4}/h.

Availability. Availability represents the percentage of time at any 24-h interval in which it is predicted that a 95% position error (due to errors in the space segment and in the control segment) is less than the threshold value for any point in the coverage area. It is based on a 95% threshold value for allowable errors in the horizontal and vertical planes and operation within the volume of service during any 24-h interval. The availability of service assumes the worst combination of two non-operating satellites.

The onboard avionics shall provide availability not worse than 0.95. For the "fail-safe" user receiver, it was determined in [13] that the availability of the GLONASS standard accuracy channel is at least 99% for servicing in the horizontal and vertical planes on the average and at least 90% for the worst sites.

Integrity. The integrity in aviation applications means the measure of confidence that can be attributed to the correctness of information output by the system in general. Integrity includes the capability of the system to provide the user with timely and reasonable alerts (alarms).

Regardless of the integrity monitoring method, the OSNE shall provide integrity monitoring by horizontal coordinates with the following characteristics [41]: Probability of missing an alarm signal is not more than 0.001 for one flight hour and probability of outputting a false alarm signal is not more than 10^{-5} for one flight hour.

For the "fail-safe" user receiver, it was determined in [13] that the integrity shall be not less than 1×10^{-7}/h for en route flights and in the terminal area, and at least $1–2 \times 10^{-7}$ per approach for the approach phase.

The analysis of the above requirements of aviation users for accuracy, continuity, availability, and integrity and their comparison with the potential capabilities of the GLONASS system show that these requirements can be ensured when using the GLONASS data. However, this does not mean that special measures are not

Table 1.1 GNSS performance

Name	Value		
	GLONASS	GPS	Galileo
Position accuracy in plan view (95%) (m)	5	3	4
Altitude accuracy (95%) (m)	14	12	10
Velocity accuracy (95%) (m/c)	0.3	0.02	0.02
Accuracy of UTC referencing	5 ns	3 ns	3 нс
Global availability (%)	100	100	99.8

required to improve the OSNE performance. In particular, various functional GNSS augmentations will be considered—space-, ground-, and aircraft-based ones ensuring the required OSNE performance.

The main characteristics of the global navigation satellite systems GLONASS and GPS considered above are given in Table 1.1.

The requirements to the GNSS performance in general are given in Table 1.2.

To ensure that the position error is acceptable, an alert limit is defined that is the largest position error providing a secure operation. The position error shall not exceed this limit without triggering an alert. Similar to the ILS, the system can degrade in the direction of increasing the error beyond 95%, but not exceeding the control limit. The requirements for alert limits are given in Table 1.3.

The requirement for the indicator of the navigation system integrity for an individual aircraft to support en route flights (without landing), operations in the terminal area, the initial approach phase, non-precision approaches, and departures determined at the GNSS receiver output is assumed to be 1×10^{-5}/h [41]. The signal in space transmitted by satellite navigation systems simultaneously serves a large number of aircraft flying along the route, and therefore, the consequences of the system integrity loss for the air traffic control system will be greater than when using conventional navigation aids.

In this regard, the requirements presented in Table 1.2 are more stringent than those for conventional non-aviation equipment of the users that allows a position error of up to 100 m.

For GNSS-based precision approach operations, the integrity requirements in Table 1.3 were selected in accordance with ILS requirements.

A range of values is defined for Cat. I precision approaches. The value of 4.0 m is determined by the technical requirements of the ILS system and is a conservative conclusion from these requirements.

The analysis of the presented structures and performance of various GNSSs shows that their main advantage is a global coverage area comprising all regions of flights of various AC types, independence from weather conditions and the capability to use them on any AC type.

However, the GNSS accuracy, integrity, continuity, and availability performance do not meet the requirements for landing systems and flight safety in general.

Table 1.2 Requirements to the GNSS performance

Flight phase	Accuracy		Integrity[b]	Time-to-alert[c]	Continuity[d]	Availability[e]	RNP type
	Horiz. 95%[a, c]	Vert. 95%[a, c]					
En route	3.7 km	N/A	1×10^{-7}/h	5 min	1×10^{-4}/h to 1×10^{-8}/h	0.99–0.99999	20–10
En route, terminal area	0.74 km	N/A	1×10^{-7}/h	15 s	1×10^{-4}/h to 1×10^{-8}/h	0.999–0.99999	5–1
Initial approach, npn-precision approach (NPA), departure	220 m	N/A	1×10^{-7}/h	10 s	1×10^{-4}/h to 1×10^{-8}/h	0.99–0.99999	0.5–0.3
Non-precision approach with vertical guidance (APV-I)	220 m	20 m	1–2 $\times 10^{-7}$ per approach	10 s	1–8 $\times 10^{-6}$ for any 15 s	0.99–0.99999	0.3/125
Non-precision approach with vertical guidance (APV-II)	16.0 m	8.0 m	1–2 $\times 10^{-7}$ per approach	6 s	1–8 $\times 10^{-6}$ for any 15 s	0.99–0.99999	0.03/50

(continued)

Table 1.2 (continued)

Flight phase	Accuracy		Integrity[b]	Time-to-alert[c]	Continuity[d]	Availability[e]	RNP type
	Horiz. 95%[a, c]	Vert. 95%[a, c]					
Cat. I precision approach[g]	16.0 m	6.0–4.0 m[f]	$1–2 \times 10^{-7}$ per approach	6 s	$1–8 \times 10^{-6}$ for any 15 s	0.99–0.99999	0.02/40

Notes

[a]To implement the planned operation at the lowest height above the threshold (HAT), the 95% value of the SRNS-defined position error is required. Detailed requirements are defined in Part B, and guidance material is given in para C.3.2 of Attachment D [13]

[b]The definition of the integrity requirement includes the alert boundary, depending on which it can be evaluated. The values of the alert boundary are given in Table 1.3

[c]Due to the fact that the continuity requirement for en route and terminal area flights, during initial approaches, non-precision approaches (NPA), and departure operations depends on several factors, including the intended operation, air traffic density, complexity of the airspace and availability of alternative navigation aids, value ranges are given for this requirement. A lower value represents the minimum requirements for areas with low air traffic density and a simple airspace structure. A higher value corresponds to areas with intensive traffic and a complex airspace structure

[d]For the availability requirements, a range of values is given, as these requirements depend on the operational need, which is based on several factors, including: frequency of operations, weather conditions, scale and duration of failures, availability of alternative navigation aids, radar coverage, air traffic density, and reversibility of operational procedures. Lower values of the requirements correspond to the minimum availability, when the GNSS system is used in practice, but cannot properly replace other navigation aids (non-GNSS). Higher values given for en route navigation correspond to the GNSS use as the only navigation aid in a certain area. Higher values for approach and departure operations meet the requirements for availability in airports with a high volume of air traffic, assuming that the landing and takeoff operations on several runways are interrelated, but separate operational procedures ensure the safety of the operation

[e]For requirements of operational readiness the range of values as these requirements depend on operational requirement, which is given it is based on several factors, including frequency of fulfillment of operations, weather conditions, scale and duration of failures, operational readiness of alternate means of navigation, radar action area, intensity of air traffic and convertibility operational procedures. Lower values of requirements correspond to the minimum operational readiness, at which system GNSS is used in practice, but cannot adequately replace other means of navigation (not GNSS). Higher given values for to routing navigation correspond to GNSS use as the unique means of navigation in some area. Higher the given values for operations of landing approach and departure meet the requirements to operational readiness at the airports about the large intensity of air traffic in the assumption that operations of landing and takeoff on several runways are inter connected, but used separate operational procedures guarantee safety of operation

[f]For precision approach the range of values is determined by category I. Value of 4, 0 m (the 13th foot) is defined by technical requirements of ILS system and is conservative conclusion from these requirements

[g]Requirements to GNSS characteristics for precision approach fulfillment on categories II and III are on consideration and will be actuated later

Table 1.3 Alert boundary values

Standard operations	Lateral alert limit	Vertical alert limit	RNP type
En route	7.4 km	N/A	20–10
En route	3.7 km	N/A	2–5
En route, terminal area	1.85 km	N/A	1
NPA (non-precision approach)	556 m	N/A	0.5–0.3
APV-I	556 m	50 m	0.3/125
APV-II	40.0 m	20.0 m	0.03/50
Precision approach (Cat. I per ICAO)	40.0 m	15.0–10.0 m	0.02/40

Currently, the GNSS itself, without auxiliary facilities and augmentation systems, is often unable to meet the requirements for the accuracy, integrity, and continuity of the data required to support approaches in accordance with the requirements for the ICAO Category I LSs [13], not to mention higher categories.

To overcome these GNSS drawbacks and for the purpose of their use by aviation users and to improve flight safety, GNSS augmentations came into common use in the late 1990s.

Next, consider the functions and structure of the main types of GNSS augmentations.

1.3 Augmentations—The Main Method to Improve the Performance Characteristics of Global Navigation Satellite Systems

To improve the characteristics of the GNSS navigation parameters generated on the basis of SV signals, special GNSS augmentations (Aug) are being developed [13, 34, 35, 42–46], which are intended for:

assessment of the compliance of the SV radiation characteristics with the standards and monitoring the navigation field characteristics;
generation of differential corrections and their transmission to the user together with the results of the navigation field monitoring;
use as additional sources of navigational information (pseudosatellites).

The reasons for the development and implementation of Aug systems are drawbacks of GNSS systems that prevent their direct use by users (including aviation ones), which require high accuracy, integrity, and continuity of navigation support.

It is known [34] that a lot of factors influence the accuracy of determining the primary navigation parameters by the GNSS receiver. They are related to the peculiarities of the navigational measurements, to the characteristics of the signals used, the propagation environment, etc. The main sources of errors in determining the primary navigation parameters are:

errors introduced on SVs or the command and measuring complex (CMC) (mainly ephemeris errors),
errors introduced on the signal propagation path (in the ionosphere, troposphere, re-reflections);
errors introduced in the GNSS receiver (mostly noise errors).

The total value of the error in determining the navigation parameters can reach several tens of meters, especially in conditions of reduced SV visibility at the location of the antenna on the aircraft, which occurs when maneuvers are performed or when flying under the conditions of hazardous terrain.

The general ideology of the construction of satellite landing systems (SLSs) on the basis of ground-based GNSS augmentations is based on the use of the concept of differential subsystems [13, 34–36] and consists in the following. On the local monitoring and correcting station located in the vicinity of the aerodrome, GNSS SV signals are received and differential corrections (DC) are calculated for pseudoranges measured on the LAAS. The DC calculation is ensured by the fact that the coordinates of the location of the antennas of the GNSS receivers that are part of the LAAS are known with high accuracy. Then, these DCs, along with the parameters of the landing glide path, are transmitted to the aircraft and used in the onboard GNSS receiver to exclude strongly correlated errors from the pseudorange measurements. As a result, the user onboard the aircraft receives the refined (updated) coordinates and time when the guidance parameters ("ILS-like" signals) are formed while performing a precision approach.

The effectiveness of the differential method depends on the degree of spatial and temporal correlation of the errors on the LAAS and on the aircraft. With a strong correlation, the systematic part of the error will be eliminated almost completely, and with a weak one, a residual error will appear, which will determine the characteristics of the navigation and landing information when performing approaches and landings.

Let us consider the basic augmentation systems and the principles of their operation. At present, it is customary to distinguish between wide-area, regional, local, and stand-alone subsystems [13, 34–36]. This classification is based on the augmentation location: For wide-area and regional Aug systems, the transmitting stations are located on SVs that are either on geostationary, or highly elliptical, or medium-altitude circular orbits. The transmitting stations of local and some regional augmentations are located on the ground. Stand-alone augmentations are built using onboard aircraft systems, such as inertial navigation systems (INSs), and are used to improve the integrity and continuity of the GNSS navigation data. Stand-alone augmentations are not considered in this paper. The operation principle of all augmentations is based on the relative constancy of a significant part of the GNSS errors in time and space, which makes it possible to evaluate these errors using independent surveillance channels, and then transmit them to the user so that his equipment could implement a differential mode of operation and generate information on the integrity of navigation parameters. The conducted studies [34, 47–51] show that the use of the differential mode makes it possible to reduce the positioning error to values less than 1 m when using code measurements and to values less than 0.1 m when using phase measurements.

Examples of wide-area augmentations (SBAS in ICAO terminology) are the American WAAS system, European EGNOS, Japanese MSAS or QZSS (Quasi-Zenith Satellite System), and Indian IRNSS. They all use geostationary SVs to transmit differential corrections and integrity information to their users. In addition, SVs of these systems can emit their own navigation signals tied to the GNSS system time and thus serve as an additional source of navigation information. Currently, a wide-area system of differential correction and monitoring (SDCM) is created in Russia, the SVs of which are located on highly elliptical orbits.

The listed systems of wide-area augmentations are designed to increase the GNSS integrity, reliability, availability, and accuracy to the level that meets the requirements of all phases of AC flights, up to the Cat. I approach.

The expected accuracy of the coordinates using SBAS is in the range of 2.5–5.0 m [44].

The systems of regional augmentations (GRAS) for aviation applications are practically not developed at present. The main focus is on the development of local augmentations (GBAS in ICAO terminology) that are planned in the future for the inclusion into the general network of SBASs and GRASs. Local augmentations have a maximum coverage of about 50 km and most often include one local monitoring and correcting station (LAAS).

A drastic solution to the problem of increasing the safety during landings is currently possible on the basis of satellite-based landing systems (SLSs) including a GNSS with LAAS-based augmentations. The main advantages of such systems are:

- relatively low cost of the ground and onboard equipment;
- possibility of accommodating the ground equipment at any aerodromes;
- low operating costs for the ground equipment;
- low cost of aircraft equipment;
- possibility of using standard aircraft antennas, which are already used in the ILS onboard equipment;
- support of a categorized landing for all runways (RWYs) of the aerodrome located within a radius of up to 50 km from the LAAS;
- flexibility, which ensures the implementation of curved approach trajectories, minimizing flight time and noise level in populated areas and landing intervals.

The analysis of the SLS characteristics shows that their implementation opens up realistic prospects for equipping the majority of airfields and aircraft, and, as a result, for increasing flight safety. SLSs were initially developed in such a way as to maintain well-established methods of conventional instrument landing systems to the maximum extent possible. This concerned both the display of information on airplane indicators and the output of signals to the automatic control system (ACS).

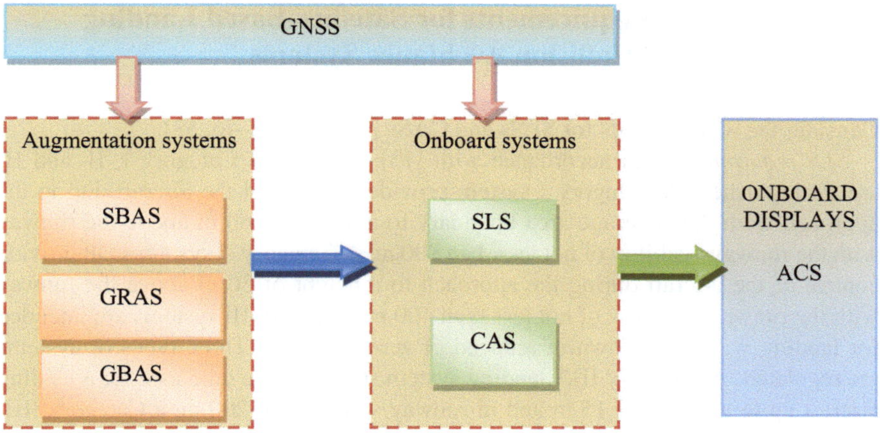

Fig. 1.6 Diagram of interaction between GNSS augmentation systems and onboard systems

In practice, when implementing an SLS, a pilot or an aircraft ACS shall not feel the difference when it comes to the question which system is used for the approach. Examples of aviation GNSS augmentations are the SLS-4000 (USA) and LAAS-A-2000 (Russia) systems [52]. An alternative LAAS based on the use of pseudosatellites takes precedence over other GBAS-like augmentations, since it does not require a special channel for transmitting corrective information and is an additional navigation point (as a result, accuracy and continuity increase). When several pseudosatellites are deployed in the terminal area, this augmentation is able to support the operation in critical situations of GNSS constellation SV malfunctions or in the case of radio countermeasures taken.

There are many options for building a ground-based GNSS augmentation including those using SRNS systems as the data transmission channel [53].

Figure 1.6 presents the generalized diagram of interaction between GNSS augmentation systems and onboard systems.

It should be noted that many onboard GNSS receivers that are used in the SLSs and CASs are capable of handling signals from several augmentations, which provide additional redundancy in integrated systems.

To address the issue of improving flight safety with the use of SLSs and CASs, it is necessary to analyze the requirements for these systems, determine which functional elements have the greatest impact on efficiency and safety, and then develop methods, techniques, and devices to solve this problem.

1.4 Analysis of Requirements for Satellite-Based Landing Systems and Collision Avoidance Systems

Consider the requirements for SLSs and CASs [3, 4, 13, 42–46, 53].

SLS requirements. In accordance with [13], systems of Category I, II, and III are distinguished: A Category I system provides control of the aircraft during the approach from the coverage area boundary to a height of 60 m above the runway with the runway visibility of not less than 800 m. A Category II system shall provide control of the aircraft during the approach to a height of 30 m above the runway with the runway visibility of not less than 400 m. Category III systems are intended for landing with touchdown in reduced or zero visibility. Three types of systems are regulated: A Category IIIA landing system shall provide guidance for a landing aircraft up to a height of 15 m and in runway visibility of 200 m, a Category IIIB system—up to touching the runway—and in runway visibility of 50 m, a Category IIIC system—with zero visibility.

Table 1.4 summarizes the general requirements for the performance of instrument landing systems using the GNSS.

These data are obtained by conservative deduction from the requirements for instrument landing systems [13] regardless of their type and, in general, provide a generalized idea of the required performance of landing systems of any type.

To improve the accuracy of navigation parameters, differential methods [28, 54] are traditionally used that are implemented using GNSS augmentations.

All differential methods are based on the use of monitoring and correcting stations (MCSs). However, the task of calculating differential corrections is in itself complex in conditions of multipath propagation of radio waves, leading to their distortion [20].

The analysis of the GNSS integrity requirements [45, 46] showed that from the total error budget (integrity loss probability) of 2.0×10^{-7} per approach, the share of SV signals (errors in the assumed pseudoranges) is 1.4×10^{-7} per approach, the share of the reference receivers of the local MCS (LAAS) and the errors related to the geometric factor is 0.5×10^{-7} per approach, the share of the onboard equipment is only 1.0×10^{-8}. This includes errors in receiving the signal of the data receiver from the LAAS and all other errors in the operation of the SLS equipment. These parameters of requirements shall be provided by a set of measures, which shall include both integrity monitoring algorithms and a priori high indicators of equipment reliability.

In order to use the GNSS under the conditions of Cat. II and III approaches, even greater accuracy and integrity is required than for Category I systems.

The LAAS provides two types of services [42, 45]: approach support and positioning. Guidance is provided on final approach segments for precision approaches, as well as for non-precision approaches (NPA) or uncategorized approaches with vertical guidance (APV) within the LAAS operational area.

The ground and onboard SLS subsystems (LAAS and GNSS/LAAS) provide position information in a horizontal plane to support area navigation (RNAV) oper-

Table 1.4 Requirements for satellite-based landing systems

Category	Error, 95%		Integrity	Continuity	Availability	Time-to-alert (s)
	Horizontal (m)	Vertical (m)				
I	16.0	4.0	$1–2 \times 10^{-7}$ (per approach)	$1–8 \times 10^{-6}$ (for 15 s)	0.99–0.99999	2
II	6.9	2.0	$1–1 \times 10^{-9}$ (for 15 s)	$1–4 \times 10^{-6}$ (for 15 s)	0.99–0.99999	1
IIIA	6.9	2.0	$1–1 \times 10^{-9}$ (for 15 s)	$1–4 \times 10^{-6}$ (for 15 s)	0.99–0.99999	1
IIIB	6.2	2.0	[a]$1–1 \times 10^{-9}$ (for 30 s) [b]$1–1 \times 10^{-9}$ (for 15 s)	[a]$1–2 \times 10^{-6}$ (for 30 s) [b]$1–2 \times 10^{-6}$ (for 15 s)	0.99–0.99999	1
IIIC[c]	3.5	1.0	[a]$1–5 \times 10^{-10}$ (for 30 s) [b]$1–5 \times 10^{-10}$ (for 15 s)	[a]$1–5 \times 10^{-7}$ (for 30 s) [b]$1–5 \times 10^{-7}$ (for 15 s)	0.99–0.999999	0.5

Notes [a]horizontal, [b]vertical, [c]draft

ations within the service area. The fundamental difference between the precision approach and RNAV functions is the different operational requirements associated with specific operations, including different data integrity requirements.

There are several possible LAAS configurations that meet the international standards [13, 45]. Among them are the following:

(a) configuration that provides only Category I precision approaches;
(b) configuration that provides Category I precision approaches and also transmits additional parameters of the position error bounds in the ephemeris;
(c) configuration that provides Category I precision approaches and positioning, while transmitting the parameters of the position error bounds in the ephemeris.

LAAS technical characteristics (accuracy, integrity, continuity, and availability of the information output) are regulated by a number of regulatory documents [13, 45]. Let us analyze these requirements.

The accuracy of determining differential corrections (DCs) shall be not worse than 0.35 m (root-mean-square error), and the accuracy of the DC rate—not worse than 0.05 m/s. The LAAS estimation by this parameter requires a standard with a known algorithm for the DC formation. Usually, the LAAS accuracy is estimated by the final result—the accuracy of the user's positioning in the differential mode (vertical and lateral deviations). It is appropriate to use this accuracy in the algorithm for monitoring the integrity of formed DCs within the LAAS itself (this procedure is called "sigma monitoring" [46]).

The alert threshold for DCs shall be no more than 0.4 m for GPS and 0.8 m for GLONASS. This parameter also applies to the integrity monitoring algorithm. Integrity, continuity, and time-to-alert parameters are set according to the required navigation performance of the current required navigation performance (RNP) specification [46].

The availability parameter characterizing the reliability of the LAAS elements shall not be worse than 1.0×10^{-4}. It should be noted that the GNSS navigation receiver interface shall ensure isolation of the functions of receivers and other LAAS equipment.

The requirements for the Cat. I approach mode (type RNP: 0.02/40) are given in [42, 45]:

- accuracy (95%)—by height: 4.4 m, lateral deviation: 9.0 m;
- error alert threshold (alarm): 10.3 m;
- maximum alert (alarm) output delay: 3 s;
- integrity risk 2.0×10^{-7} per approach (150 s).

The ground subsystem (LAAS) shall provide the following:

- integrity risk: 1.5×10^{-7};
- continuity risk: 8.0×10^{-6} for 15 s (of them 1.1×10^{-6} are for the data transmission channel, and 6.9×10^{-6} for the LAAS, while the greater part of the LAAS risk is for the loss of satellite tracking (4.5×10^{-6}) and integrity tests (2.3×10^{-6});
- availability of 99.9% (availability risk of 10^{-3}).

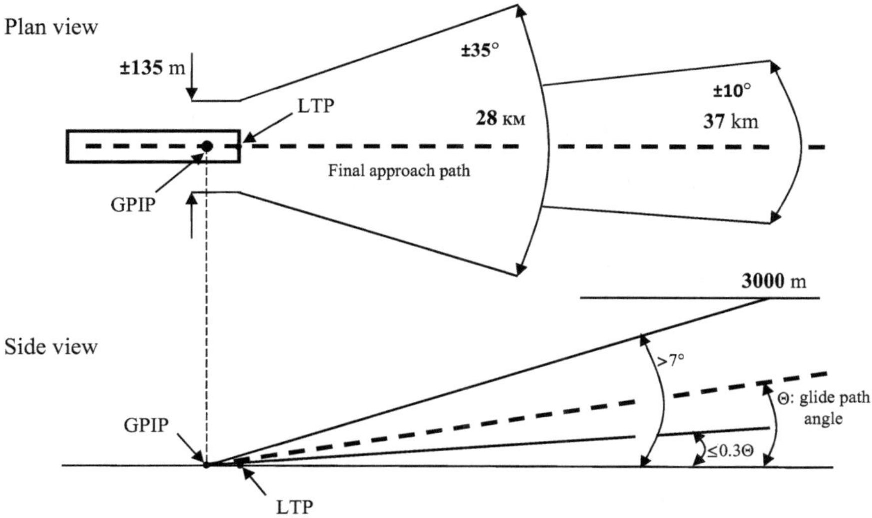

Fig. 1.7 LAAS minimum coverage area

The requirements for integrity and continuity are for ensuring a higher accuracy than specified above. In addition, these requirements may vary depending on the application of the system. For example, the requirements for continuity risk depend on the airspace structure in the landing area (air traffic density, airspace complexity, availability of alternative navigation aids, etc.). Accordingly, the requirements for integrity, continuity, and availability may be more stringent in future (integrity risk of 1.0–2.0 × 10^{-7}, continuity risk of 1.0–8.0 × 10^{-6}, availability of 99.9–99.999%). At the same time, the requirements for accuracy will also become higher.

The minimum LAAS coverage area to support approaches is shown in Fig. 1.7.

When the LAAS transmits additional parameters of the position error bounds in the ephemeris, the differential corrections can only be used within the maximum use distance (D_{max}) defined in the LAAS type 2 message [13].

The LCMS coverage area, where the aircraft positioning is determined using the GBAS, depends on the planned specific operations.

The optimum coverage area for this service shall be omnidirectional in order to ensure the aircraft positioning outside the coverage area for the precision approach.

For precision approaches and operations with SLS-based positioning (GBAS), various levels of integrity are determined.

The risk of the signal-in-space integrity loss for a Category I landing is 2.0 × 10^{-7} per one approach, the duration of which usually does not exceed 150 s. This value can be visualized as follows: It is acceptable not to output an alert for the pilot indicating that the parameters of the navigation system are out of tolerance, in no more than two cases out of 10 million approaches. The SLS determines errors in the corrected pseudorange error relative to the LAAS reference point (σ_{pr_gnd}), as

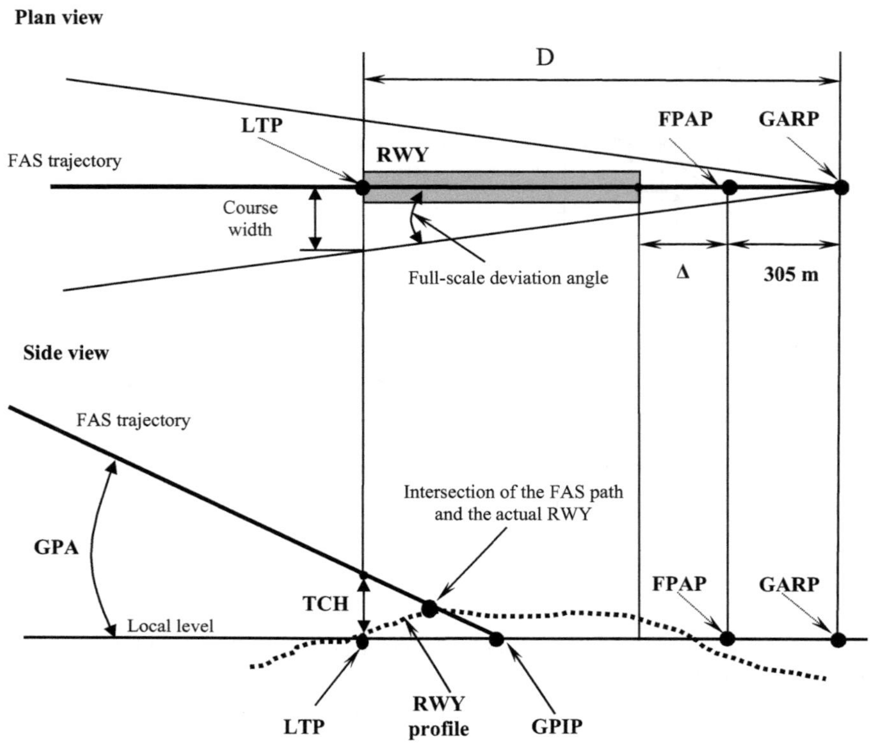

Fig. 1.8 Definition of the FAS path

well as the errors due to the vertical (σ_{tropo}) and horizontal (σ_{iono}) decorrelation due to the spatial diversity of the LAAS and the aircraft. These errors are used in the GNSS/LAAS onboard equipment [42] to calculate errors in the navigation solution and to generate an integrity violation alert if the error exceeds acceptable thresholds.

In the final approach phase, the SLS usually uses a trajectory, which is a line in the space defined by the runway threshold (LTP/FTP), the flight path alignment point (FPAP), the threshold crossing height (TCH) and the glide path angle (GPA). These parameters are determined from the information contained in the FAS data block in the LAAS type 4 message or in the onboard database [13]. The relationship between these parameters and the FAS path is illustrated in Fig. 1.8.

The Δ-length offset determines the distance from the runway end to the FPAP. This parameter is introduced so that the onboard equipment could calculate the distance to the runway end.

The local approach vertical is defined as the normal to the WGS-84 ellipsoid at the LTP/FTP and can differ significantly from the local gravity vector. The local horizontal approach plane is defined as a plane perpendicular to the local vertical passing through the LTP/FTP (i.e., tangent to the ellipsoid at the LTP/FTP). The datum crossing point (DCP) is a point at the TCH height above the LTP/FTP. The FAS path is defined as a line passing through the DCP and located at an angle equal to the glide path angle (GPA) relative to the local horizontal plane. The GPIP is a point at which the final approach path crosses the local horizontal plane. The GPIP can actually be above or below the runway surface depending on the runway curvature. To provide compatibility with existing types of landing systems (ILS, MLS), the onboard GNSS/LAAS equipment shall provide guidance information in the form of deviations with respect to the desired flight path defined by the FAS path. Such deviations in satellite-based landing systems are called "ILS-like."

For various purposes, various ways of calculating deviations can be used. One of the calculation methods is presented in [55]. The standard algorithm is defined in [13, 42].

In view of the strict requirements for the accuracy of the aircraft positioning during Category I approaches, the FAS parameters shall be determined with high accuracy: The error in specifying the LTP/FTP and FPAP coordinates does not exceed 0.015 m. The TCH value is set with a resolution of 0.1 m, and the glide path angle (GPA) is set with increments of 0.01°.

The LAAS VDB transmitter is an important part of the LAAS and provides emission of data and corrections to the GNSS ranging signals via VHF broadcasting of digital data in the frequency range of 108–118 MHz with channel division of 25 kHz. This range makes it possible to use the ILS LOC receiver antenna onboard the aircraft as an antenna for the onboard VDB receiver.

Given the high requirements for the data integrity performance during precision approaches, the LAAS transmitter is subject to rather stringent technical requirements. For example, the carrier frequency stability is maintained in the range of ±0.0002% of the allocated frequency; data is coded by a bit-by-bit phase shift. The messages themselves are formed as symbols, each of which consists of three consecutive bits of the message, and the symbols are converted into a differential eight-level format (D8PSK) through a phase shift of the carrier frequency by 45°. The symbol rate is maintained at 10,500 symbols/s (±0.005%) and provides a nominal data transmission rate of 31,500 bps [13, 42].

Since the desired coverage area for the SLS-based positioning can be larger than the coverage area of one LCMS, a network of LCMSs can be used to provide it. These stations can transmit signals on one frequency and use various time intervals (currently eight intervals are defined) at adjacent stations to avoid interference, or they can transmit at different frequencies.

The VDB receiver shall have high noise immunity and provide interference rejection on the working channel in the presence of signals from such systems as VOR, ILS LOC, and VHF FM broadcasting signals.

To ensure all types of paths for aircraft traffic in the terminal area, the radiation pattern of the VDB receiver AFS in the horizontal plane shall be omnidirectional.

The lateral and vertical alert limits [13] for Category I precision approaches are calculated in the SLS in accordance with Table 1.5 [13].

The vertical alert limit for Category I precision approaches is measured from an altitude of 60 m above the LTP/FTP. For a procedure using a decision height of

Table 1.5 Alert limits in Category I SLSs

Horizontal distance (D) from the aircraft to the LTP/FTP (m)	Lateral alert limit (m)
$0 \le D \le 873$	FASLAL
$873 < D \le 7500$	$0.0044\,D$ + FASLAL − 3.85
$D > 7500$	FASLAL + 29.15
AC height (H) above the LTP/FTP (m)	Vertical alert limits (m)
$H \le 60$	FASVAL
$60 < H \le 410$	$0.095965\,H$ + FASVAL −5.85
$H > 410$	FASVAL + 33035

more than 60 m, the VAL at this decision height will be larger than the transmitted FASVAL parameter.

If a fault is detected and if there is no backup transmitter, the LCMS service shall be terminated if the signal cannot be reliably used in the coverage area. To meet the requirements for errors in determining the AC navigation parameters during approaches, the following requirements are imposed on the relative accuracy of the geodetic survey of the LAAS reference point [45]:

- The error in the geodetic survey of the LAAS reference point relative to the WGS-84 coordinate system shall be no more than 0.25 m vertically and 1.0 m horizontally.
- For each LAAS reference receiver (RR), the error of the reference antenna phase center fix shall be no more than 0.08 m relative to the LAAS reference point.

Moreover, it is recommended to provide even smaller errors in the determination of the indicated LCMS parameters for improving the SLS operational performance in general.

CAS requirements. In view of the fact that the requirements for collision avoidance systems are formulated as requirements for outputting alerts in case of situations when the parameters of the AC state are in the phase space zones, which presents a potential collision hazard, the most important parameter for the CAS is accuracy of the navigation information sources and the accuracy of digital databases of terrain data and artificial obstacles.

Figure 1.9 (the left part) provides examples of phase planes with alert zones, where the CAS generates the corresponding alerts when the aircraft state parameters are in these zones [56].

The right part shows illustrations explaining the meaning of the alert that occurs.

Figure 1.10 (the left part) is an example of the configuration of protective spaces in the CAS used to generate alerts in the mode of forward terrain assessment. The right part is a picture explaining the cause of the alert.

The main source of information for CAS modes considered in this paper is the GNSS.

The regulatory CAS documentation states [29] that the GNSS accuracy is sufficient to support the CAS operation.

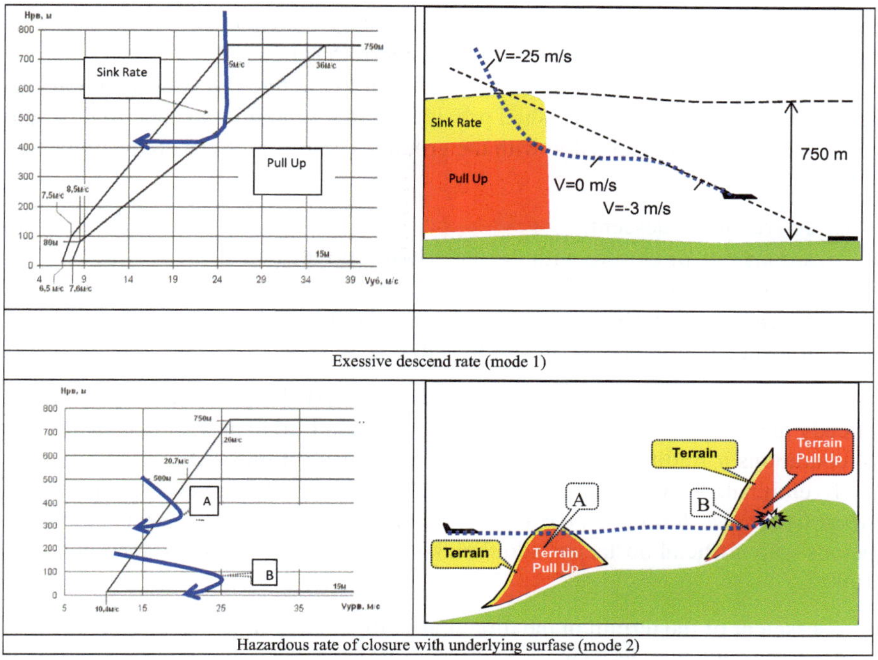

Fig. 1.9 Examples of phase planes and alert zones in the CAS

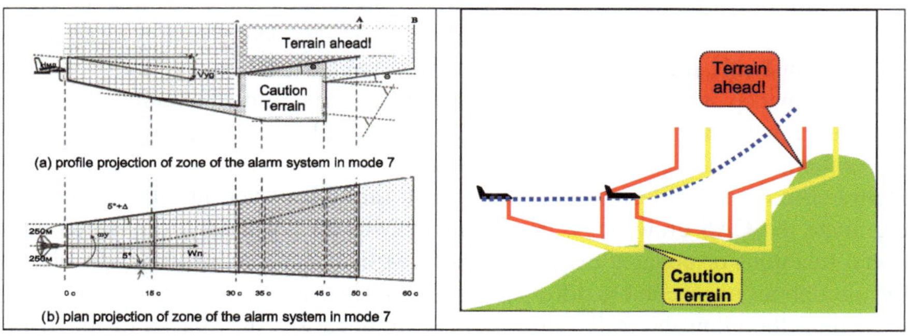

Fig. 1.10 Configurations of CAS protective spaces in the forward terrain assessing mode

Any increase in the accuracy, integrity, and continuity of the definition of navigation parameters that can be achieved by improving the performance of the GNSS radio navigation information (also with augmentation systems) will generally increase flight safety.

One of the sources of errors in the CAS is inaccuracies of digital terrain relief databases, the requirements for which are determined by documents [57].

In the present paper, these errors will not be considered.

It should only be noted that errors in determining, for example, the terrain height currently do not exceed 1 m for flights in the terminal area [58].

Regulatory documents [5, 28, 32, 34, 54, 59–67] define the basic functions and requirements for CASs, which shall provide alerts for the pilot in such flight conditions that can lead to a collision with the ground. These conditions are formulated as follows:

- excessive rate of descent;
- hazardous rate of closure with the underlying surface;
- loss of altitude after takeoff;
- approaching the underlying surface in a non-landing configuration;
- significant deviation below the glide path line.

The conditions listed above are intended to alert the pilot that the aircraft is already in conditions, the development of which can lead to an accident. This group of conditions is traditionally called the "main modes" of the CAS.

To eliminate the CAS drawbacks caused by the late output of alerts, in particular, when flying over hazardous terrain and when the aircraft is not protected from collisions with artificial obstacles, additional new functions were introduced in the late 1990s:

- assessment of the terrain in the direction of flight (forward terrain assessment);
- alerts of premature descent;
- generation and output of the underlying surface image to the display [29].

The introduction of these requirements was largely due to the fact that the "main modes" of the system no longer provided the necessary safety requirements in the face of constantly increasing flight intensity, as well as the emergence of GNSSs, the performance of which made it possible to begin mass implementation of systems with new functions. This new group of functions is called the "early warning mode."

As it was mentioned earlier, the emergence of CASs and their further improvement led to a certain decrease in the accident rate. Moreover, in order to partially eliminate the CAS drawbacks, various manufacturers constantly introduced additional functions or updated existing ones. In particular, such functions appeared as the alert about unacceptable differences in barometric and radio altimeter readings, exceeding the allowable roll value, etc. To eliminate the drawbacks of the "early warning mode" function group, various methods were used to form the predicted flight path and to reduce the false alarm rate described, for example, in [28, 56].

A common drawback of all the CAS functions implemented so far is that they actually warn of an event that has already occurred or of the fact that this event may soon occur. This makes the pilot take actions to solve the hazardous situation, stopping the implementation of the current mission or tasks of the current flight phase. At the same time, a fundamentally important problem is the development of such methods that would prevent the aircraft from ending up in a state that requires additional recovery actions from the crew, and for UAVs, it is important to generate such control signals that will prevent an emergency situation.

As part of this paper, it will be shown how the main methods aimed at enhancing flight safety by improving the accuracy, integrity, and continuity of the GNSS radio navigation signals lead to increased efficiency in the aviation use. Obviously, this is due to the use of universal GNSS technologies applied both in civil and military areas.

Now, having analyzed the main characteristics of the SLSs and CASs, we turn to the consideration of methods and techniques, the implementation of which will solve the problem of increasing flight efficiency and safety.

1.5 General Methods and Techniques to Improve Flight Efficiency and Safety When Using Satellite-Based Landing Systems and Collision Avoidance Systems

Despite the expected high economic and technical efficiency of SLS and CAS implementation using GNSS technologies, a number of fundamental technical problems need to be overcome to ensure their practical implementation. The main goal pursued is to ensure compliance with the requirements for landing systems in general.

One can distinguish several basic problems the methods of solution for which will be considered below. Such problems include: rational choice of the GNSS receiver; ensuring the accuracy, integrity, continuity, and availability of the GNSS navigation information; support of the system operation in the event of GNSS degradation or in conditions of interference; problems of preventing aircraft from ending up in hazardous situations both in flights under the conditions of hazardous terrain and during landings.

The most important element of the GNSS hardware, from the perspective of its use in SLSs and CASs, is receivers, which determine the technical characteristics of these systems to the considerable extent. The use of the GNSS as a basis for building SLSs and CASs involves the use of GNSS augmentation systems and the choice of a specific type of receiver for the rational construction of their structures.

Let us consider in more detail the main content of the methods and techniques of improving flight safety, which are presented in Fig. 1.11 and are investigated in this paper.

At first glance, the rational choice of a GNSS receiver seems to be a fairly simple task. However, as will be shown in Sect. 1.2 of this paper, such a choice belongs to the class of multiple factor problems, for the solution of which it is necessary to develop appropriate methods, including those using the fuzzy sets theory. Wrong choices lead to inefficient development and operation of the system.

If the GNSS receiver is correctly selected, the SLS and CAS may improve the quality of navigation information, including increasing the accuracy of the aircraft positioning, integrity, continuity, and availability of the data. These characteristics determine the probability of successful landing and, therefore, flight safety in general.

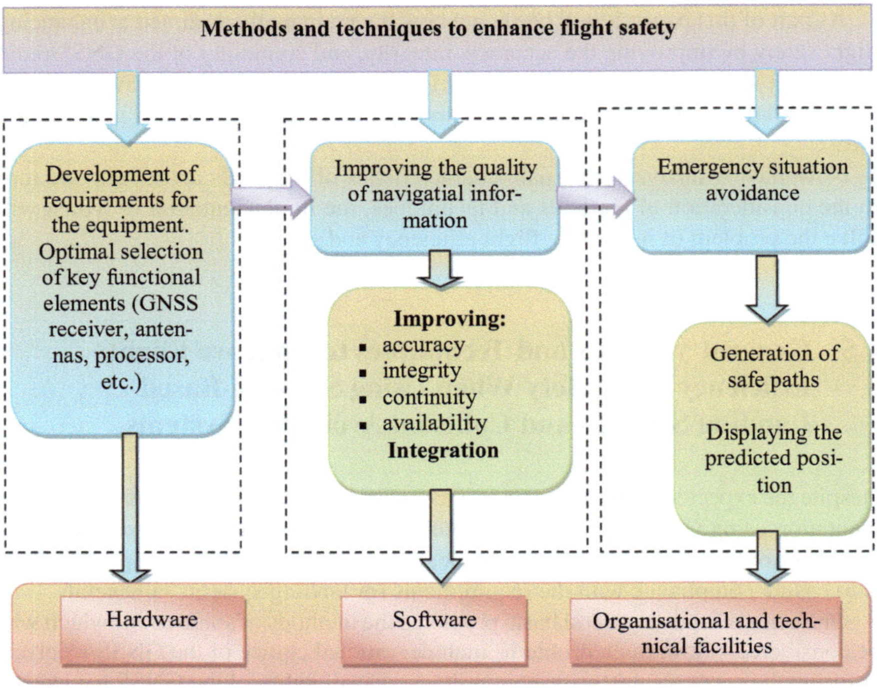

Fig. 1.11 Methods and techniques to enhance flight safety

Avoidance of emergency (hazardous) situations, which is the main objective of the CAS, can be implemented by various methods. In this paper, we consider the most general methodological approaches to the generation of safe paths and displaying the predicted aircraft position during the landing.

1.6 Flight Safety Indicators

Flight safety (FS) has always been a decisive factor in all aviation activities. This is reflected in the goals and objectives of the International Civil Aviation Organization (ICAO) [5, 59–64].

In any system, it is necessary to define and measure the final indicators in order to determine the conformity of this system to the expected results and to identify possible areas where certain measures are required to improve the results to achieve the expected level.

ICAO adopted the concept of an acceptable FS level that meets the need to use an approach based on safety indicators, specified FS levels, and FS requirements, for which it is possible to establish reliability and/or accuracy indicators.

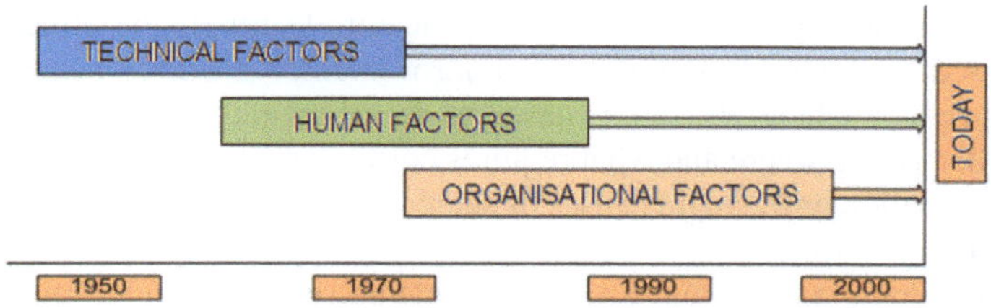

Fig. 1.12 Evolution in accounting for factors affecting flight safety

At present, ICAO determines the necessary condition for reducing the probability of accidents below 10^{-6} (i.e., one accident with human casualties per million flights) [5].

An example of the safety level can be expressed in the following way: 0.5 accidents with human casualties per 100,000 flight hours for airlines (safety indicator), with a 40% reduction in this coefficient over 5 years (target safety level).

In the traditional approach, the main focus of the FS provision was on compliance with increasingly more complex regulatory requirements. This approach was quite effective until the late 1970s, when the dynamics of the accidents (A) became even. Accidents continued to occur, despite all the rules and regulations. This approach to flight safety included retroactive actions against undesirable events by imposing measures aimed at preventing their recurrence. Instead of identifying best practices or desirable standards, efforts focused on ensuring compliance with minimum standards. With a frequency of accidents with human casualties of about 10^{-6}, which is currently the indicator that ICAO requires from the aviation community, further improving the FS level with this approach became increasingly difficult.

A modern approach that provides an acceptable level of risk in the face of expanding aviation activities is to move from a pure response to a proactive approach [68]. It is believed that in addition to a strong legislative base and regulatory requirements based on the requirements of ICAO Technical Standards (SARPs), an effective role in the FS management is played by a number of factors, such as the use of evidence-based methods for managing risk factors, providing comprehensive training of operations personnel in the FS area (including human factor aspects), etc.

None of the elements of the FS improvement individually can meet today's expectations regarding the management of risk factors. Only the integrated application of most of these elements can increase the stability of the aviation system to unsafe actions and conditions.

In general, ICAO's approach to FS provision can be presented in the form of increasing risk management accounting for various factors, as shown in Fig. 1.12.

There is no single FS indicator that would be acceptable for all cases [68]. The indicator expressing a target safety level shall be appropriate for the application area, in order to allow an effective assessment of the safety performance using the same parameters that were used to determine the specified safety level.

Before proceeding to the selection of indicators to assess the effectiveness of the proposed methods of efficiency FS improvement, we briefly consider the FS performance as a whole, the main causes of accidents and high-risk flight phases. This will determine where and what requires correction or improvement first of all.

The distribution of the main causes of accidents according to the ICAO statistics is shown in Fig. 1.13.

The most common accident causes are weaknesses in preflight training, inadequate pilot qualifications, pilot errors in estimating distance and speed, inadequate knowledge of aircraft features by the crew, continuation of flights in adverse weather conditions, and inadequate assessment of the descent rate during approaches by pilots.

Figure 1.14 provides the statistics of accidents over last years in Russia, the USA, and the world as a whole [63].

Figure 1.15 shows accident factors in passenger transportation by airplanes with takeoff weights exceeding 10 ton in the CIS, in the world in general and in the USA [63, 69–71].

According to Boeing [70], in civil aviation from 1959 to 1995, 56.7% of all accidents in the world took place during initial and final approaches and landings, 20.1% during takeoffs and initial climbs. The cruise flight phase accounts for 5.7% of accidents.

Table 1.6 shows a comparison of the distribution of accidents by flight phases for world civil aviation (CA), military transport aviation (MTA), and long-range aviation (LRA) [28].

Despite the measures taken by the aviation community, the share of accidents attributed to the so-called human factor remains high (60–70%) and practically does not decrease [5]. This indicates a partial exhaustion of the possibilities of such activities aimed mainly at improving the human operator performance.

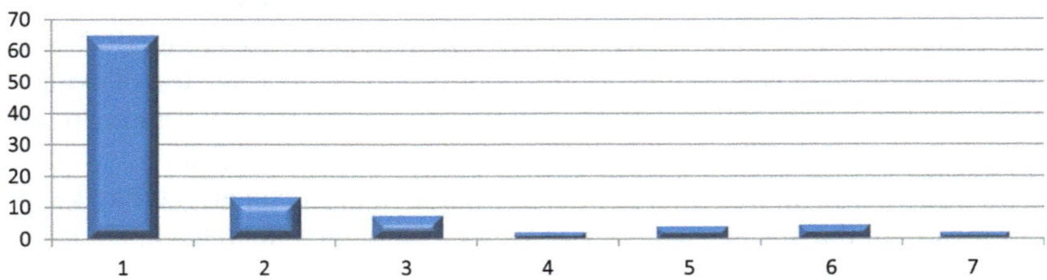

1 – flight technical error and crew's incorrect in–flight decision (65%); 2 – unsatisfactory flight control, violation of flight operations rules by the crew (13.6%), 3 – erroneous actions of crews (7.6%); 4 – shortcomings in the organization of the customer's work (2.3%); 5 – event causes not identified (4.1%); 6 – shortcomings in the ATC and meteorological services work (4.6%); 7 – shortcomings in the airport services work (2.2%).

Fig. 1.13 Main causes of accidents

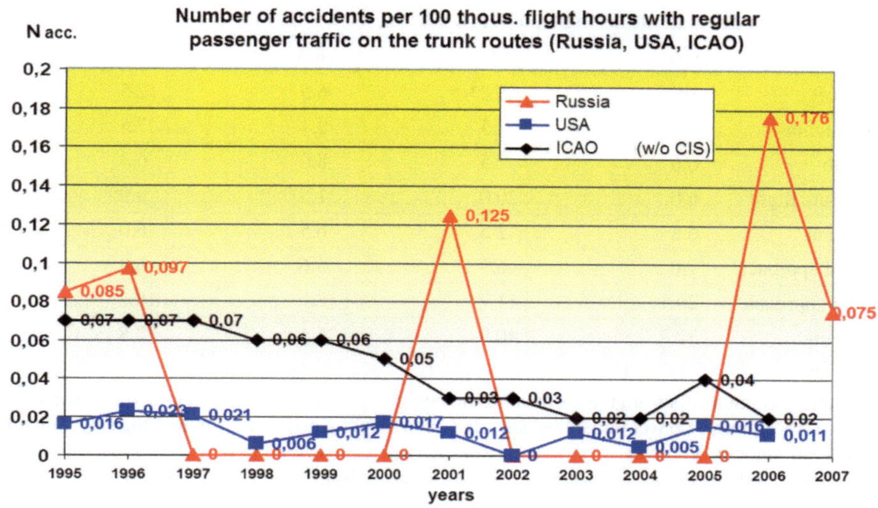

Fig. 1.14 Flight safety indicators

Another direction to reduce the accident rate is the modernization of the aircraft design. Thus, the ICAO Accident Prevention Manual states that the aircraft design shall provide for a reduction in the probability of human errors.

In other words, the machine shall "forgive" human errors and mitigate their consequences.

If errors in themselves are not obvious, then the crew shall receive a signal indicating their appearance [57].

The analysis of the data presented above and of the ICAO safety concept allows for the following conclusions:

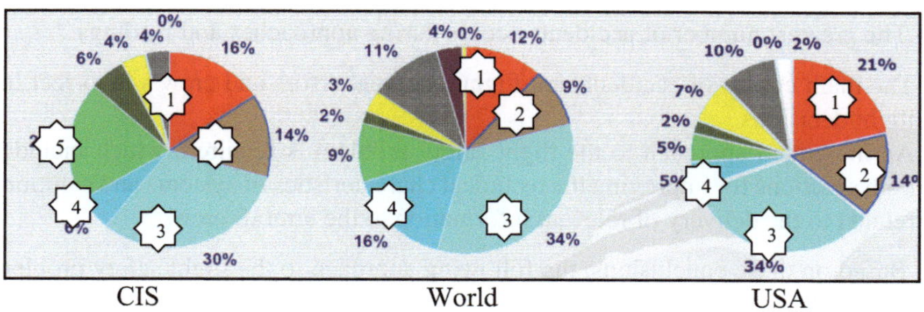

Equipment failures (1). Crew errors in case of equipment failures (2). Crew errors (3). Crew errors and adverse external effects (4). Crew and ATC errors (5). ATC drawbacks (6). Adverse external effects (6). Diversions, air defense (7) Unidentified causes (8). Other causes (9).

Fig. 1.15 Accident factors in passenger transportation by airplanes with takeoff wights of more than 10 ton

Table 1.6 Accidents at various flight phases (B %)

Flight phase	CA	MTA	LRA	Average
Takeoff	14	8.23	6.5	12.5
Initial climb	7.4	7.5	8.1	7.5
Climb	6.6	2.5	8.0	6.3
En route flight	6.0	20.0	31.5	9.9
Descent	6.2	2.5	6.5	6.0
Initial approach	7.0	6.7	5.6	6.8
Final approach	20.1	11.7	8.9	18.2
Landing	32.7	40.8	25.1	32.8

Notes

The *takeoff* phase starts from the moment of the beginning of the movement along the runway and ends with reaching a conventional obstacle clearance height equal to 10.7 m and the safe value of the takeoff velocity Vtakeoff

The *initial climb* begins at an altitude of 10.7 m and extends up to an altitude of 400 m

The *climb* begins at the altitude of the circle and ends with reaching an intended flight level and cruising speed

During the *en route flight*, the airplane flies at an established flight level, which is one of the surfaces of constant atmospheric pressure

At the *descent* phase, the airplane flies with a loss of altitude along an inclined path

At the *initial approach* segment, a pre-landing maneuver is performed along a racetrack pattern, during which the airplane performs four turns

The *final approach* phase begins from the moment the airplane crosses the glide path. At the intersection of the glide path, the airplane enters a descent mode along the inclined glide path

The *landing* phase starts at a height of 15–20 m and, in the most general case, consists of four sections: leveling, holding, sinking, and rolling. In the process of leveling, the airplane moves along a curved path, which, by the end of this segment, makes it possible to reduce the vertical speed to a safe value. At the holding segment, the airplane flies horizontally at an altitude of 0.5–1 m in order to reduce the horizontal speed to the allowable value. When the airplane is sinking, it moves along a curved path until the ground contact. After landing, the airplane begins the final phase of landing—rolling that ends with a complete stop of the airplane

The greatest number of accidents occurs during approaches and landings.

- The main causes of accidents are flight technical errors and crew's incorrect in-flight decisions.
- An integrated approach to the flight safety problem is required, which includes both solutions for improving the technical characteristics of systems and solutions related to the delivery of relevant information to the aircraft crew.

Based on these conclusions, the following approach to the flight safety problem is proposed:

1. It is necessary to develop and implement a landing system that will be capable of servicing all types of aircraft in any region of the globe and which has high accuracy, integrity, and continuity in determining navigation parameters. As such

a system, at present, the only alternative is the satellite-based landing system (SLS).
2. In order to avoid incorrect actions of the crew, it is necessary to develop and implement such a collision avoidance system (CAS) that will prevent incorrect actions by the crew at various flight phases, including high-risk phases: landing and rolling. As such a system, it is advisable to use modern collision avoidance systems based on the use of GNSS technologies.
3. In view of the fact that the GNSS is the functional core of both SLSs and CASs, it is possible to find such ways and principles of joint (integrated) use of these systems, which will significantly improve flight safety.

We now introduce a group of indicators that will allow assessment of the effectiveness of the proposed approaches to FS improvement and reduce the probability of accidents.

At present, statistical indicators are used to assess the FS level. These indicators are common and partial, absolute and relative [5]. Common indicators (accidents, catastrophes, casualties, etc.) characterize flight safety as a whole. Particular indicators (the number of accidents for the *j*th cause or at the *i*th phase) describe specific causes or at certain phases. More universal are relative FS indicators, in which the number of events with aircraft is given for a certain amount of operation time or works performed [68]:

$$K_f^i = n_f^i / L_f, \tag{1.6}$$

where n_f^i is the accident number, L_f is the amount of works (operation time), f is an event type index, and i is an aircraft type index.

Statistical FS indicators are objective criteria and in this lies their core value. At the same time, these criteria also have a number of significant drawbacks: They assess flight safety after an accident took place; they cannot be used to predict the FS level, since they do not take into account the features of the new equipment, changes in the conditions of its operation, do not allow to determine the degree of hazard of adverse factors and their influence on flight safety and, therefore, cannot be used to find effective ways to prevent accidents before their practical implementation.

The main factors that affect flight safety and determine the occurrence of special situations (accidents) in flight are: external disturbances affecting the aircraft and arising with probability $P_1(t)$, equipment failures with probability $P_2(t)$, incorrect actions of the crew when piloting with probability $P_3(t)$, etc. Each of the in-flight affecting factors can be represented as a random process with a finite set of states $S_n\{S_i\}$, $(i = 1, m)$ and a continuous region of variation in the flight parameters x_i. Such a random process is described in general form by the Markov model of the performance function $S_n\{S_1, S_2, \ldots, S_m\}$ [68] in the form of a probability vector $P_{AF}(t)$

$$P_{\mathrm{AF}}(t) = \left[P_1(t, S_1), P_2(t, S_2), \ \ldots, \ P_j\left(t, S_j\right), \ \ldots\right] \tag{1.7}$$

of the onboard navigation and landing complex (ONLC) being in any of the states S_j.

The flight process from the ONLC preflight check to the landing is usually divided into n phases, with A_k events of the successful completion of the kth phase being considered independent. When each kth phase is performed, the known jth combination of ONLC systems is used, and the fact B_{jk} of successful operating of the jth combination of systems involved at the kth flight phase is represented as event S_{ljk}, when each lth ONLC system in the jth combination at the kth flight phase performs its functions (i.e., is operable). The hazard level of the current flight mode in the event of a malfunction of the ONLC systems can be estimated by the probability P_{Acc} of the occurrence of an accident type Q in the form [68]:

$$P_{\text{Acc}}(t) = \left[\prod_{k=1}^{n}\sum_{j=1}^{m}\prod_{l=1}^{p} P\big(S_{ljk}\big)P\big(H_i/S_{jlk}\big)P(Q/H_i)\right] \times \left[\prod_{k=1}^{n}\sum_{j=1}^{M}\prod_{l=1}^{N} P\big(S_{ijk}\big)P\big(H_i/S_{ijk}\big)P(Q/H_i)\right], \tag{1.8}$$

where M is the number of possible combinations of technical systems at the kth flight phase; N is number of auxiliary technical systems in combination r at the kth flight phase; $P(S_{ljk})$ and $P(S_{qrk})$ are the probability of failure for the lth ONLC element in the jth combination at the kth flight phase and the failure of the qth auxiliary system supporting the kth flight phase in the combination r; $P\big(H_i/S_{ljk}\big)$ and $P\big(H_i/S_{qrk}\big)$ are conditional probabilities of accident type H_i with the failure of the lth ONLC element and the qth auxiliary system in the corresponding combinations at the kth flight phases.

The analysis of the expression (1.8) for the calculation of P_{Acc} parameter shows that even if it is simplified for one flight phase (landing) and only two systems (SLS and CAS) are considered, then to obtain the final result giving an idea of the achieved FS level, many limitations and assumptions are required, which in general is not the purpose of this study.

Therefore, in the present paper we will use indicators that characterize the relative increase in the factors that affect flight safety as a whole, for each specific proposed method, and as an initial level, we will consider the technical level achieved so far characterized by a specific parameter.

In view of the fact that this paper considers two radio technical systems that can be used both separately and jointly, the assessment of the safety of the integrated system will involve the classical Bayesian formulas for estimating a priori and posterior probabilities [72–75]. For example, the conditional probability of the accident occurrence in the joint (integrated) use of the two systems will be determined by the product of the accident probability when using one system (P_{SLS}) and the accident probability when using the other system (P_{CAS}).

For convenience in analyzing the results obtained, we introduce the following indicators (relative factors):

- reduction in error ($\mathbf{F_A}$), which is the ratio of the positioning before and after application of the proposed methods for increasing accuracy;
- reduction in the accidents probability ($\mathbf{P_{Acc}}$), which is the ratio of the parameters defining the boundaries and areas of the phase spaces used to generate alerts before and after application of the proposed methods;
- increase in integrity ($\mathbf{F_{Int}}$) and continuity ($\mathbf{F_{Cont}}$), representing the ration of relative calculated values of the relevant indicators before and after application of the proposed methods;
- increase in ergonomics ($\mathbf{F_{Erg}}$), which is the ratio of the time estimated for the pilot's decision to perform maneuvers during flights under the conditions of hazardous terrain or when landing on an unequipped aerodrome before and after application of the proposed methods; the same factor will characterize the relative decrease in the false alarm probability;
- integration ($\mathbf{F_{Integration}}$), which is the ratio of the conditional probabilities of performing a safe flight when one of the systems is used and when they are used together.

In the presented statement of the problem, an analytical solution is impossible because of the absence of functional dependencies of P_{Acc} on F_j, ($j = A$, Acc, Int, Cont, Erg, Integration), and there is no functional relationship between the flight safety indicators and the accident probability [68]. In this paper, as the main indicator characterizing the effectiveness of the developed methods and devices, we will consider the relative increase in the probability of performing a safe flight, which we will characterize by the set of factors presented above.

If we represent the total probability of performing a safe flight P_{FS} as a generalized functional, which is a logical generalization of expression (1.8), and write it in the form:

$$P_{FS}(t) = F\left(S_{1,}S_2, \ \ldots, \ S_n, \ \ldots\right), \tag{1.9}$$

where S_i functions depend on the above factors $\mathbf{F_A}, \mathbf{P_{Acc}}, \mathbf{F_{Int}}, \mathbf{F_{Cont}}, \mathbf{F_{Erg}}, \mathbf{F_{Integration}}$: $Si = S_i\ \varphi(F_j)$, *where* $j = A$, Acc, Int, Cont, Erg, Integration, then the only representation suitable for practical use is its decomposition into a sum of partial probabilities proportional to the chosen factors. Then, taking into account the chosen indicators, the problem under consideration can be represented by the following expression:

$$P_{FS}(t) = P_{FS0}(t) + \sum_j D_j \cdot F_j, \tag{1.10}$$

where $P_{FS}(t)$ is the probability of safe flight *using* the proposed methods and devices, $P_{FS0}(t)$ is the probability of safe flight *without using* the proposed methods and devices, D_j is the contribution of a factor to the overall probability of safe flight, F_j is the estimate of the corresponding factor, index j takes the values: {A, Acc, Int, Cont,

Erg, Integration}. The dependence of the probability of safe flight on time (t) means that in the general case the considered probability is non-stationary. Taking into account the fact that the calculation of the D_j factors was not the goal of the present paper, and their magnitude is not generally determined, further only F_j factors are calculated, which characterize the relative increase in the probability of safe flight or the effectiveness of the guidance systems using the developed methods and devices.

In view of the selected flight safety improvement indicators when using satellite-based landing systems and collision avoidance systems, the problem to be solved can be represented as follows:

$$P_{FS}(t) = P_{FS0}(t) + \sum_j D_j \cdot F_j \rightarrow 1. \quad (1.11)$$

Expression (1.11) shows that, in general, any increase in the relative F_j factors leads to an increase in flight safety and the more of these factors can be improved, the higher the probability of a safe flight as a whole.

1.7 Conclusions

The use of conventional instrument landing systems and the collision avoidance systems practically exhausted all possibilities of its development in terms of increasing flight safety. Reducing the probability of accidents and catastrophes below the level of 1.0×10^{-6} with the use of conventional systems becomes practically impossible or economically inefficient.

Currently, in Russia less than 5% of the airfields are equipped with instrument landing systems. The equipping of most aerodromes with conventional instrument landing systems is economically impractical. The only way to improve flight safety is to develop and implement satellite-based landing systems.

The wide introduction of the GNSS in aviation and the coverage of 100% of the Earth's territory with the GNSS navigation field having an availability level exceeding 0.99999 and accuracy of several meter provide a new approach to the problem of flight safety using instrument satellite-based landing systems and collision avoidance systems.

The development of a network of GNSS augmentation systems based on local monitoring and correcting stations (LAAS) makes it possible to reduce the error in determining navigational parameters to 1 m or less when the GNSS/LAAS onboard equipment is operating in the differential mode. This is sufficient to ensure instrument precision approaches in accordance with ICAO requirements. The use of LAASs solves the problem of increasing the integrity of navigation information to 2.0×10^{-7} per approach.

The analysis of the requirements to the performance of satellite-based landing systems and collision avoidance systems showed that it is appropriate to address the problem of improving flight safety with the approach involving the combined use of these two systems. This approach provides, among other things, the performance of a number of functions of one system based on information from the other one.

The analysis of accident (Acc) statistics shows that the largest number of accidents (up to 70%) occurs during approaches and landings. The main causes of accidents (up to 50%) are pilot errors and crew's incorrect in-flight decisions including those under the conditions of equipment failures.

Thus, in order to address the problem of improving flight safety, it is first and foremost necessary to focus on the approach and landing phase, with particular attention to ergonomics issues. A common methodology is needed for the construction of satellite-based landing systems and collision avoidance systems that will allow the development of new structures of the systems under consideration, new methods and devices for improving flight safety, and will ensure their practical implementation.

Ways and approaches to solving this problem are discussed in the following chapters.

References

1. Sosnovsky AA, Khaimovich IA (1975) Short-range radio engineering and landing systems. Mechanical Engineering, Moscow, 200 p (in Russian)
2. Sosnovsky AA, Khaimovich IA (1990) Aviation radio navigation: handbook. Transport, Moscow, 264 p (in Russian)
3. RADIONAVIGATION PLAN OF THE RUSSIAN FEDERATION. The main directions of development of radio navigation systems and aids (edition 2008) [Electronic resource]. http://stat.doc.ru/documents/ (in Russian)
4. Amendment No. 27 to Annex 6, Part 1. Convention on International Civil Aviation. ICAO, letter of recommendation AN 11/11/26-01/61 of June 15 2001 (in Russian)
5. Safety management manual [Electronic resource]. Doc 9859, First edition. International Civil Aviation Organization, 2006 (in Russian)
6. Silvestrov MM (ed) (2007) Ergonomic integrated aircraft complexes. Branch of Military Publishing House, Moscow, 512 p (in Russian)
7. Minimum Performance Standards—Airborne Ground Proximity Warning Equipment. In: RTCA DO-161A, 1976
8. Ground Proximity Warning Equipment (GPWS)/ARINC CHARACTERISTIC 723, AERONAUTICAL RADIO, INC., 1988
9. Joint requirements to the airworthiness of civil transport aircraft of the CMEA member countries (ENLG-C)/IAC for the airworthiness of civil aircraft and helicopters of the USSR, 1985. 470 p (in Russian)
10. AVIATION REGULATIONS (2003) Part 23. Airworthiness requirements for civil light aircraft. In: Interstate aviation committee. OJSC "AVIAIZDAT", Moscow (in Russian)
11. AVIATION REGULATIONS (2004) Part 25. Airworthiness requirements for transport aircraft. In: Interstate aviation committee. OJSC "AVIAIZDAT", Moscow (in Russian)
12. AVIATION REGULATIONS (1995) Part 29. Airworthiness requirements for transport rotorcraft. In: Interstate aviation committee. OJSC "AVIAIZDAT", Moscow (in Russian)
13. Annex 10 to the Convention on International Civil Aviation (2006) Aeronautical telecommunications. Radio Navigation Aids. 6th edn, vol 1. (in Russian)
14. Sauta OI, Gubkin SV (1987) Statistical characteristics of the SRNS azimuth and range information during the landing. Radioelectronics issues. Series OVR-1987-issue 7, pp 23–29 (in Russian)

15. Sauta OI (1987) Substantiation of the possibility of using domestic SRNSs during approaches. Radioelectronics issues. Series OVR-1987-Issue 5, pp 39–48 (in Russian)
16. Sauta OI Use of radio beacons of the short-range radio technical navigation system to support instrument approaches. Radioelectronics issues. Series OVR-1987-Issue 5, pp 39–48 (in Russian)
17. Baranov YuYu, Kurochkina SL, Sauta OI Adaptive differential algorithm for analysis of statistical reliability of radio technical information (in Russian)
18. Sauta OI Substantiation of the coordinate system selection in the construction of algorithms for complex information processing in the landing system based on the short-range radio technical navigation system (in Russian)
19. Ground proximity warning system. SSOS. 6G.1.700.009 RE, 1975. (in Russian)
20. Terrain Awareness and Warning System. TAWS-1. Manual for technical operation. 6G1.700.009 RE, 1984 (in Russian)
21. IL-62 aircraft (1983) Ground proximity warning system "Vector". Methodical manual of the aviation engineer, issue 2. Air Transport, Moscow (in Russian)
22. David L (1996) FSF launches final assault on 'killer' CFIT accident rate. Flight Int 15
23. David L (1994) Task force plans to halve CFIT incidents. Flight Int 1994, p 11
24. William BS (1995) New technology, training target CFIT Losses. Aviat Week Space Technol 73–77
25. Edward HP (1997) FAA May mandate enhanced GPWS. Aviat Week and Space Technol 22–23
26. Edward HP (1997) Safety of nonprecision approaches examined, Aviat Week Space Technol 23–29
27. Terrain Awareness And Warning System (2002) TS0-C151B. Department of Transportation Federal Aviation Administration, USA
28. Dryagin DM (2006) Complex enhanced ground proximity warning system with enhanced functionality and software algorithms that minimize the false alarm probability [Electronic resource]: Thesis Tech Sciences: 05.11.03. RSL, Moscow (in Russian)
29. Sauta OI (1990) To the use of the SRNS system to support approaches on under-equipped airfields. Radioelectronics issues. Series OVR-1990-Issue 5, pp 29–37 (in Russian)
30. Marine doctrine of the Russian Federation for the period until 2020. [Electronic resource http://stat.doc.mil.ru/documents] (in Russian)
31. Bakit'ko RV et al (2005) GLONASS. In: Perov FI, Harisov VN (eds) Principles of construction and functioning, 3rd edn. Radioelectronics issues, 688 p (in Russian)
32. Military doctrine of the Russian Federation for the period until 2020/Approved by the Decree of the President of the Russian Federation of 5 Feb 2010 No. 146 "On the Military Doctrine of the Russian Federation". [Electronic resource] http://stat.doc.mil.ru/documents (in Russian)
33. Baranov YuYu, Gubkin SV, Sauta OI (1989) Study of digital SRNS filters for high-speed aircraft. Radioelectronics issues. Series OVR-1989-Issue 10, pp 3–10 (in Russian)
34. Dmitriev PP et al (1993) Network satellite radio navigation systems. In: Shebshayevich VS (edn) 2nd edn. Radio and Communication, Moscow, 408 p (in Russian)
35. Solovyev YuA (2003) Satellite navigation and its applications. Eco-Trends, Moscow, 326 p (in Russian)
36. Yatsenkov VS (2005) Fundamentals of satellite navigation. NAVSTAR and GLONASS GPS systems. Goryachaya liniya-Telecom, Moscow, p 272 (in Russian)
37. Global Navigation Satellite System (2008) GLONASS. Interface control document. Version 5.1. RNII of Space Instrumentation, Moscow (in Russian)
38. Interface control document. Rev.C. ICD-GPS-200, ARINC, 2000
39. Performance-based navigation (PBN) manual. Third edition, ICAO, 2008 (in Russian)
40. Qualification requirements "Onboard equipment for satellite navigation" QR-34–01. Edition 4. Interstate Aviation Committee, 2011 (in Russian)
41. Qualification requirements "GNSS/LAAS onboard equipment" QR-253. Edition 1. Interstate Aviation Committee, 2007 (in Russian)
42. Qualification requirements "GNSS/SBAS onboard equipment" QE-229. Revision 1. Interstate Aviation Committee, 2011 (in Russian)

43. RTCA DO-209 (2006) Minimum operational performance standards for the Global Positioning System/Wide Area Augmentation System Airbone equipment. Radio Technical Commission for Aeronautics
44. RTCA DO-245A (2004) Minimum aviation system performance standards for local area augmentation system (LAAS). Radio Technical Commission for Aeronautics
45. RTCA DO-246C (2005) GNSS based precision approach local area augmentation system (LAAS)—signal-in-space interface control document (ICD) [Electronic resource]. Radio Technical Commission for Aeronautics
46. Sokolov AA (2005) Estimation of the LAAS differential correction errors in aviation. Abstract of thesis … and Tech Sciences. S.-Pb (in Russian)
47. Sokolov AI (2006) Monitoring phase measurements in reference receivers of ground local segmentation systems GPS/Glonass. In: Sokolov AI, Chistyakova SS (eds) Publication House SPbGETU "LETI"/SPbGETU "LETI", 2006.-Issue 2. Radioelectronics and Telecommunications, pp 27–32 (in Russian)
48. Chistyakova SS (2007) Study of the effect of phase measurements jumps on errors in determining the aircraft coordinates in the instrument differential satellite-based landing system. In: 62nd scientific and technical conference on the radio day. Proceedings of the conference. Apr 2007. St. Petersburg. Publ. House SPBGETU "LETI" (in Russian)
49. Sauta OI, Sauta AO, Chistyakova SS, Yurchenko YuS, Sokolov AI, Sharypov AA (2011) Effect of the multipath signal propagation on measurement errors in the global navigation satellite system. In: Fourth all-Russian conference "fundamental and applied position, navigation and time support" (PNT-2011). Theses of reports. Institute of Applied Astronomy, Russian Academy of Sciences, St. Petersburg, pp 224–226 (in Russian)
50. Sauta OI, Sokolov AI, Yurchenko YuS (2009) Estimation of data generation accuracy in the local area augmentation system in aviation. Radioelectronics issues. Series RLT-2009-Issue 2, pp 183–193 (in Russian)
51. Integrated complex (in Russian)
52. Local Area Augmentation System. Certificate of type № 399 of 02.12.2005. Interstate Aviation Committee. Aviation register (in Russian)
53. Baburov VI, Rogova AA, Sobolev SP (2005) Method for calculating deviations from the heading line and glide path in the onboard equipment of the satellite-based landing system. NSTU Sci Bull 1:3–10 (in Russian)
54. Certificate of the product conformance. SGKI-034-112-SRPBZ. Interstate Aviation Committee. Aviation register, 2003 (in Russian)
55. Instruction manual. Enhanced ground proximity warning system. EGPWS. RShPI.461531.001 RE. VNIIRA, 2004 (in Russian)
56. Standards for Processing Aeronautical Data. RTCA DO-200A. Radio Technical Commission for Aeronautics, 2005
57. Annex 4 to the Convention on International Civil Aviation. Aeronautical charts, 11th edn, July 2009 (in Russian)
58. Myasnikov EV (2000) High-precision weapons and strategic balance. In: Center for the study of disarmament, energy and environmental problems at MIPT, Dolgoprudny, 43 p (in Russian)
59. Annual report of the Council. 2011 [Electronic resource]. International Civil Aviation Organization. Doc 9975, www.icao.int
60. Aralov GD (2012) Analysis of flight safety issues. Flight safety issues, №5. VNIITI, Moscow (in Russian)
61. Annual report of the Interstate Aviation Committee (IAC) "Flight Safety Status in 2011" [Electronic resource], www.mak.ru (in Russian)
62. Materials of the 5th international conference "Safety of the air transport system" [Electronic resource]. Moscow, 20 Feb 2012 (in Russian)
63. Guziy AG, Lushkin AM (2008) Quantitative estimation of the current level of aircraft operator flight safety. Flight safety issues, No. 10. VNIITI, Moscow (in Russian)
64. Abstract of a study conducted by the European Aviation Safety Agency (EASA) for 2009. Flight safety issues, No. 11, 2010. VNIITI, Moscow (in Russian)

65. Terrain Awareness And Warning System (TAWS). ARINC 763, Recommendations dated 10 Dec 1999 (in Russian)
66. Akos D (2000) Development and testing of the stanford LAAS ground facility prototype. In: Akos D, Gleason S, Enge P, Luo M, Pervan B, Pullen S, Xie G, Yang J, Zhang J (eds) NTM 2000: Proceedings of 2000 National technical meeting of the institute of navigation. Anaheim, 2000, pp 210–219
67. Braff R (2001) LAAS performance for terminal area navigation. In: ION 57th annual meeting/CIGTF 20th biennial guidance test symposium, 11–13 June 2001, Albuquerque, NM, pp 252–262
68. Reports of the Interstate Aviation Committee on the flight safety status in civil aviation in 1992–2004. IAC, Moscow, 231 p (in Russian)
69. World Aircraft Accident Summary—CAA CAP479, Airclaims, London TW6 2AS, 1963–2004
70. Compliance with ICAO Standards—Key to Flight Safety. Flight Safety, 25 Mar 2002, pp 12–19 (in Russian)
71. ICAO Accident Prevention Manual, Doc 9422—AN/923
72. Bronshtein IN, Semendyaev KA (1965) Handbook on mathematics. Moscow Nauka (in Russian)
73. Vygodsky MYa (1973) Handbook on higher mathematics. Nauka, Moscow (in Russian)
74. Korn G, Korn T (1974) Handbook on mathematics for scientists and engineers. Nauka, Moscow (in Russian)
75. Baburov VI, Ivantsevich NV, Sauta OI (2011) Formalized approach to the selection of a receiver for onboard satellite navigation and landing equipment. Radioelectronics issues. Series OT.-2011-Issue 4, pp 16–31 (in Russian)

Chapter 2
Methodology for Constructing Satellite-Based Landing Systems and Collision Avoidance Systems

Satellite-based landing systems (SLSs) and collision avoidance systems (CASs) are systems of integrated radioelectronic complexes (RECs), the functional core of which is a GNSS receiver that generates navigation parameters and precise time.

The methodology for constructing SLSs and CASs involves the development of the main stages of selecting individual elements, on the basis of which systems are developed, and then the selection and formation of structures in which methods, techniques, and devices are implemented to solve the problems of FS increasing.

When developing structures of the systems under consideration, a whole spectrum of problems arises; and solutions of them determine not only their technical characteristics, but also the amount of financial costs and time frame required for their costs [1, 2]. The wrong choice of REC key elements leads to significant losses of resources at the development stage and to a decrease in the efficiency of the system during the operation phase [3].

The main methodological stages of the SLS and CAS development are presented in Fig. 2.1.

The development begins with an analysis of technical requirements (specifications) for systems (stage I), which involves identifying the main and auxiliary elements necessary to construct the system as a whole.

In view of the fact that the development is based on GNSS technologies, at the same stage, the basic GNSS elements shall be identified, which will be used in the system, including GNSS augmentation systems.

At the same stage, an optimum GNSS receiver shall be selected, which is the SLS and CAS functional core.

At stage II, it is reasonable to consider methods for ensuring the basic requirements for the systems being created, since they determine the structure of the systems and all the necessary connections between the various elements within these structures.

At stage III, a synthesis of the SLS and CAS structure is made, on the basis of which perspective elements for integration and complexing of functional, hardware and software features are determined.

At stage IV, the effectiveness of the proposed solutions is assessed by analyzing the generalized indicator of flight safety improvement P_{FS} (see Chap. 1).

© Springer Nature Singapore Pte Ltd. 2020

Baburov S.V. et al., *Development of Navigation Technology for Flight Safety*, Springer Aerospace Technology, https://doi.org/10.1007/978-981-13-8375-5_2

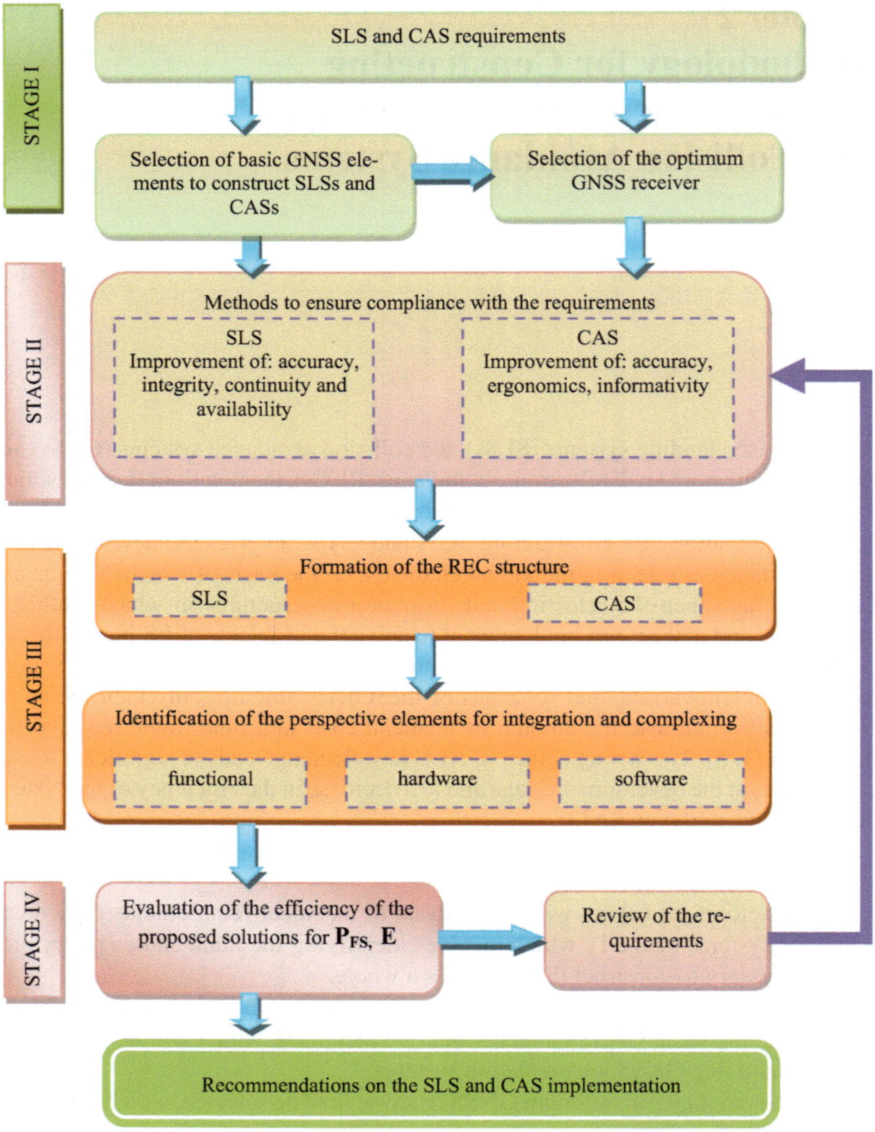

Fig. 2.1 Main methodological stages of the SLS and CAS development

If the result obtained corresponds to the expected result, then the proposed methods can be implemented in SLSs and CASs.

Otherwise, it is necessary, for example, to review the requirements and repeat steps II and III (and possibly I) again.

This section deals with stages I and partially III of the proposed methodological approach, including the methodology for selecting a built-in GNSS receiver and an illustration of the results of its use on a reduced set of alternatives and preference criteria. Then, the SLS and CAS structures are reviewed, where functional elements are singled out, for which implementation methods are needed to be developed, which is the goal of the approach to FS improvement.

The results presented in this chapter are described in more detail in [1–5].

2.1 Theoretical Background of a Formalized Methodological Approach to the Selection of Basic Elements for Radioelectronic Complexes to Improve Flight Efficiency and Safety

The main point of the approach proposed below is a multistage gradual narrowing of the set of alternatives, first using a reference matrix and clear (quantitative) characteristics of alternatives, and then by analyzing the remaining alternatives using fuzzy sets theory with fuzzy (qualitative) characteristics. The integral indicator of quality is the generalization of the results of the analysis of clear characteristics, and the use of a unified approach based on the fuzzy sets theory is the generalization of the proposed approach.

Figure 2.2 presents a generalized diagram of the proposed approach.

Let us consider in more detail the actions performed during the implementation of the proposed approach in accordance with Fig. 2.2, taking the GNSS receiver as an example.

The set of functional and operational parameters of GNSS receivers used in the multiple factor analysis includes several hundreds of indicators grouped by their types. Samples of these parameters are conditionally divided into type groups. Group I contains mandatory requirements, usually included in the terms of reference (ToR) for the system. When solving optimization problems, they are transferred to the category of constraints. Group II includes additional requirements, which make it possible to expand the system functions. Such parameters, for example, describe the possibility of interacting with a GNSS receiver (availability of "feedback" with the manufacturer, the possibility to implement additional options, etc.), can take into account the previous experience of using the products of certain manufacturing companies, the capability of products to expand their functionality, etc.

For example, Table 2.1 shows the main parameters of GNSS receivers related to group I, and Table 2.2, respectively, those related to group II.

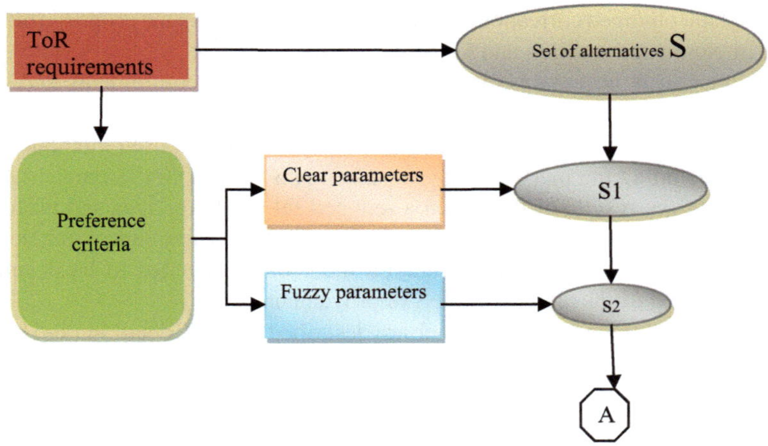

Fig. 2.2 Generalized diagram of the approach to the selection of alternatives

Table 2.1 Mandatory requirements

Requirements	Description
Constructional	
Dimensions	
Weight	
Electrical energy consumption	Power voltage, consumpted power
Reliability	Mean time between failures (MTBF)
Resistance to external (environmental) effects (EEs)	Temperature, vibration, shock, etc.
Interface	Availability and number of standard interfaces
Functional	
Errors in positioning, velocity and time	In standard and differential modes
Frequency of data output without extrapolation	5 Hz
Operational frequency ranges	L1
Time characteristics	"Hot" and "cold" start times, re-acquisition time
Interference immunity	
Sensitivity and input signal dynamic range	
Tracking system dynamics	Speed, acceleration, jerk of the intended operating facility
Built-in test system	RAIM, FDE
Economic	
Cost	
Production and delivery time	

Table 2.2 Additional requirements

Requirements	Description
Constructional	
Electrical energy consumption	Protection, control, power for the active antenna, non-volatile memory, integrated battery (accumulator)
Module structure	Cardboard for self-contained unit, set of microcircuit chips, etc.
Functional	
Operational frequency ranges	L2, L2C, L5, E5, etc.
Radiopath parameter control	
Navigation solutions	Reception of messages from pseudosatellites
Correlator	Gate width control, synchronization from the external reference generator
Errors in positioning, velocity and time	Standard mode (GLONASS, GPS, GALILEO), differential modes (SBAS, GRAS, GBAS)
Attitude parameters	Heading, roll, pitch
Frequency of data output without extrapolation	20 Hz min
"Raw" data	Navigation messages, satellite coordinates
Signal phase processing	Data smoothing with phase measurements
Interface	Special
Radiopath channels	Universal, dedicated
Software	Selection of the frequency plan, tuning to new signals, user reset, possible correction on demand
Built-in test system	Built-in simulator
Service tasks	En route navigation, relative navigation, dead-reckoning, LAAS selection, path recording
Interference immunity	Reject filter, special algorithms, etc.
Antenna	Number of antenna inputs, possible antenna parameter control
Internal memory	Volume, type
Reference frames and databases	Ellipsoids, geoid, magnetic variations
Reliability	Additional elements of protection
Antenna	Electric power, control
Economic	
Availability of control	Input–output
Warranties	From supplier, manufacturer

Consider the preference criteria in solving the GNSS receiver selection problem, which optimizes the procedure for their search from an existing set. These criteria, as well as the functional and operational parameters of the GNSS receiver, are not always quantifiable or described in a clearly mathematical way. A suitable apparatus for solving optimization problems in selecting a particular receiver using the preference criteria are methods based on the fuzzy sets theory [1]. The apparatus of the fuzzy sets theory provides a good result in the multiple factor analysis, where the use of other methods is often impossible. The use of this apparatus when analyzing the intersections of groups of preference criteria and groups of GNSS receiver characteristics makes it possible to optimize the procedure for search or ordering a GNSS receiver, speeding up the development process of the system as a whole.

From the set of alternatives S (existing receivers), further, we will analyze only a subset of S1 alternatives, for which the requirements of group I are met (see Fig. 2.2).

Consider the preference criteria of group II. For the SLS and CAS development, they can be ordered as follows: (F_1) integrability into the structure of the system, (F_2) data reliability and integrity, (F_3) time for implementation, (F_4) cost, (F_5) measurement errors. Therefore, henceforth, we will consider a unified methodological approach.

Depending on the specific objectives of the developer, it is possible to apply other criteria. To simplify the presentation of the results, we restrict ourselves to the above five criteria.

Each of the j preference criteria Fj is characterized by a set of $n(j)$ parameters, of which $n_1(j)$ parameters a clear quantitative description, and $n_2(j)$ are qualitative characteristics that allow only a fuzzy description. For example, for criterion F_1, we distinguish the following concepts that are quantifiable: weight–dimensions, power consumption, resistance to external effects (EEs), and number of interface channels. The qualitative characteristics of the F_1 criterion are availability of expert advice and the possibility of making adjustments to the functioning algorithms.

Thus, for the example under consideration, we obtain: $n(1) = 6$, $n_1(1) = 4$, $n_2(1) = 2$.

Similarly, a list of clear and fuzzy characteristics is compiled for each criterion F_j.

For the data reliability and integrity criterion $F_{2,}$ such parameters are used as the probability of delivery of reliable information to the user in a preset time interval. Integrity is usually characterized by such quantitative indicators as the probability of missing an alarm signal, the probability of outputting a false alarm signal, the alarm signal delay time. It should be noted that when creating a particular system, the criterion F_2 is decisive, because integrity and reliability indicators are currently the main characteristics of the system as a whole [1]. As indicators of the F_2 criterion reliability, the following is usually used: mean time between failures (MTBF), probability of the product malfunction detection by the built-in test system.

The F_3 criterion (time required to build the receiver into the system) shall take into account the time of the complete development cycle for a particular product. For example, when using a receiver with characteristics already known (estimated in other products or complexes), time savings can be achieved by reducing the amount of tests that must be performed with a receiver whose characteristics are not fully known.

When using the F_4 criterion (cost), it is necessary to consider the characteristics of the manufacturer of a particular receiver. Termination of the production of the selected receiver may entail such temporary and material losses that will greatly downweigh the initial benefit from minimizing the receiver cost. These characteristics are qualitative and allow only an unclear description.

When using the F_5 criterion (errors in measuring navigation parameters), it should be taken into account that at present the quantitative indicators characterizing this criterion are very similar for the vast majority of receivers. If for a particular receiver the error differs by more than 50%, then this indicates to the fact that in general this product has extremely low consumer properties. To simplify the presentation of the main results of the selection, we restrict ourselves to the above five criteria.

At the first stage of the analysis for the selected criteria (F_1–F_5), we consider characteristics that are quantifiable. For a comparative analysis of these characteristics, we construct a matrix of weighting factors, with which we estimate the degree of proximity of each alternative (particular receiver) to the "ideal" receiver (IR).

The IR concept considers such a receiver, the integration of which into the system structure does not require hardware and software modifications and which corresponds to all ToR restrictions. From the avionics developer's perspective (SLS, CAS, etc.), the IR is a finished product, the use of which requires a minimum of time and costs to solve a particular problem. The IR characteristics are formed according to the results of expert assessments. The IR concept is used as a reference standard against which the characteristics of receivers actually available in the market are compared.

The goal of optimization in the problem under consideration is the selection of such a receiver that performs the functions assigned to it in the best way and provides an optimal design solution in accordance with the criteria considered above.

The general approach to solving the problem of selecting a GNSS receiver involves several steps. We shall consider the analysis in the simplest (linear) case, when the mutual influence of the criteria is not taken into account and only quantitative characteristics are considered.

To evaluate the proximity of each alternative (receiver) from the subset S1 to the IR, we construct a matrix $\mathbf{P_s}$ with the dimension $[j \times m]$, where $m = \max(n_1(j)), j = 1$–5. The elements of the matrix $\mathbf{P_s}$ determine the quantitative characteristics of a particular receiver or are zero if they are not present in the description of the F_j criterion.

For the IR, we construct a reference matrix $\mathbf{N}$, the structure of which is similar to the matrix $\mathbf{P_s}$. The elements of the matrix $\mathbf{N}$ are inverses of the numerical values of the expert estimates of the IR parameters for the F_j criteria under consideration.

Then, the generalized index I_s of the proximity degree of a particular receiver to the ideal RI can be represented in the form:

$$I_s = \left[(\mathrm{diag}(T \cdot P_S \cdot N^T) \cdot J\right]^{-1}, \tag{2.1}$$

where **T** is a diagonal matrix of dimension [$j \times j$] with elements $t_i = 1/n_i(i)$, $j =$ 1–5, and elements of the vector **J** are expertly assigned weights for the considered F_j criteria, which are chosen from the condition:

$$\sum_{i=1}^{j} j_i = 1, \quad j_i \geq 0.. \tag{2.2}$$

In general, the distribution of the criteria weights given by the vector **J** can be arbitrary within the specified constraint. If it is necessary to narrow the search field, zero weights can be assigned to some of the criteria, which exclude the corresponding parameters from the analysis of alternatives.

It follows from (2.1) that if the IR is analyzed, then the generalized indicator $I_s = I_{IR} = 1$. For any other receiver, this indicator is $I_s < 1$. Thus, the strategy for selecting the best alternative I_0 in analyzing the quantitative characteristics of the receiver is as follows:

$$I_0 = \max_{s}\{I_s\}. \tag{2.3}$$

It is expedient to analyze the qualitative characteristics of alternatives according to the preference criteria F_j using methods of the fuzzy sets theory [6].

We denote by μ_j(S2, t), where $t = 1 \ldots n_2(j)$, the expert value of the membership function for the qualitative parameter within the F_j criterion on the subset of alternatives S_2 remaining after the previous stage of analysis of the quantitative characteristics of the alternatives S_1 ($S_2 \subset S_1$).

Consider the convolution of preference relationships (criteria) of the following form [6]:

$$\mu_0\{S_2, t\} = \sum_{i=1}^{S_2} \lambda_i \mu_i(S_2, t), \tag{2.4}$$

where λ_i are corresponding weighting factors selected from the condition:

$$\sum_{i=1}^{j} \lambda_i = 1, \quad \lambda_i \geq 0.$$

Using expression (2.3) allows ordering the alternatives of the subset S_2 in terms of their degree of dominance.

Let us illustrate the approach described above in solving the problem of selecting an embedded module of the GNSS receiver.

As ToR restrictions, let us consider the two most significant ones: the temperature ranges from −40 to +85 °C; power consumption of not more than 1 W.

The characteristics of the ideal receiver will be described by a reference matrix **N**, which is based on expert estimates.

Expert estimates of the quantitative parameters of the criterion $\boldsymbol{F}_1$ (integrability) are currently characterized as follows: dimensions of 4.5 × 3.5 × 4.0 cm max (corresponding to 63 cm^3 in volume), mass of 50 g max, power consumption of 0.5 W max, operating temperature ranges from −40 to +85 °C, permissible vibration of 5 g min, interface with three channels at least.

The qualitative parameters of the criterion $\boldsymbol{F}_1$ for the IR are characterized as follows: availability of expert consultation from the receiver developer is 100% (the value of the parameter is equal to one), possibility of adjusting the operation algorithms of the receiver is 100% (the parameter value is one).

For the criterion $\boldsymbol{F}_2$ (reliability and integrity), the quantitative parameters of the ideal receiver currently have the following indicators:

- probability of missing an alarm signal is 1.0×10^{-6},
- probability of outputting a false alarm signal is 10^{-5},
- alarm signal delay time is 0.2 s,
- mean time between failures (MTBF) is 50,000 h,
- probability of the malfunction detection by the built-in test system is 0.99.

The parameters of the criterion $\boldsymbol{F}_3$ (time for implementation) for the ideal receiver can be expertly estimated as follows: At present, the complete cycle of manufacturing a prototype product (from the development of design and programming documentation to the completion of preliminary tests) averages one year [1]. If the implementation of the ideal receiver in the total labor costs does not exceed one month, this parameter is sufficient from the criterion $\boldsymbol{F}_3$ perspective.

The analysis of the current market for GNSS receivers [7, 8], the use of which is potentially possible as part of satellite-based landing and collision avoidance systems, shows that the quantitative parameter of the criterion $\boldsymbol{F}_4$ (cost) for the ideal receiver can be estimated at USD 500.

The criterion $\boldsymbol{F}_5$ (measurement errors) for the ideal receiver is characterized by such an indicator as the position error (dX) with a probability of 0.95. The use of a GNSS receiver in the systems in question implies that it shall work in both standard and differential modes. At present, it can be expertly considered [7] that for the ideal receiver in the standard mode, dX can be 7 m, and in the differential mode, it does not exceed 1 m.

The above expert estimate of quantitative characteristics of the ideal receiver characterized by the matrix **N** can now be presented as follows:

$$N = \begin{vmatrix} 63 & 50 & 0.5 & 1/40 & 1/85 & 1/5 & 1/3 \\ 1 \times 10^{-6} & 1 \times 10^{-5} & 0.01 & 1/50000 & 1/0.05 & 0 & 0 \\ 1 & 0 & 0 & 0 & 0 & 0 & 0 \\ 500 & 0 & 0 & 0 & 0 & 0 & 0 \\ 7 & 1 & 0 & 0 & 0 & 0 & 0 \end{vmatrix},$$

where the numerical values indicated above are given for each of the selected preference criteria in the corresponding rows.

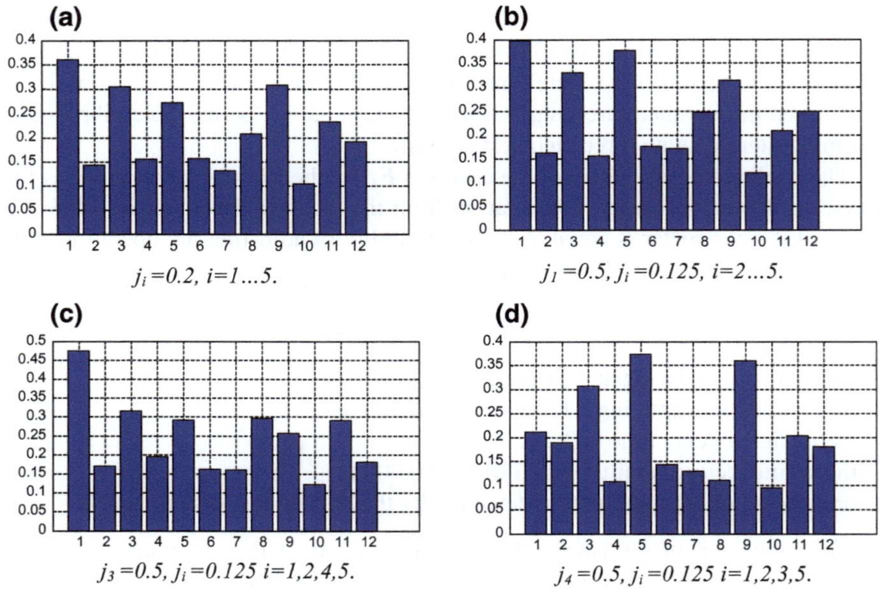

Fig. 2.3 Value of the generalized indicator I_s for various weights j_i of the criteria F_j

490	0.035	1.2	40	85	10	2
1.0-10^{-8}	10^{-7}	0.1	200 000	0.94	0	0
1	0	0	0	0	0	0
4000	0	0	0	0	0	0
4	0.2	0	0	0	0	0

Fig. 2.4 Characteristic matrix

Figure 2.3a shows the results of calculating the generalized indicator I_s for 12 alternatives (receivers) with equal weights j_i of the criteria $\boldsymbol{F}_j$ ($j_i = 0.2$). The abscissa in Fig. 2.3 depicts the conditional number of the receiver, and along the axis of ordinates, the values of the generalized indicator I_s are given.

Figure 2.3b–d shows the results of calculating the generalized indicator I_s for different weights j_i of the criteria F_j.

Figure 2.3b shows F_1 (integrability) as the dominant criterion, for which the weight $j_1 = 0.5$ is taken, and the weight of the remaining criteria is set to 0.125. Similarly, Fig. 2.3c shows F_3 (time for implementation) as the dominant criterion, and Fig. 2.3d shows the criterion F_4 (cost).

An example of the characteristic matrix $\mathbf{P_s}$ for a receiver with a conditional number 5 is shown in Fig. 2.4.

The analysis of the data shown in Fig. 2.3 shows that in the presented set of alternatives there are obvious preferences, i.e., such receivers for which the generalized indicator I_s reaches the greatest values. Moreover, the preferred alternatives (receivers) have the highest I_s values regardless of the selected criteria weights (in a fairly wide range of weights: from 0.5 to 0.125), which indicate the correctness of the set of preference criteria used and the stability of the proposed method for calculating the generalized indicator I_s.

For five alternatives (Nos. 1, 3, 5, 9, 11 in Fig. 2.3) with the largest values of the indicator I_s that are produced by Russian enterprises and currently available on the market, qualitative characteristics were evaluated using fuzzy sets methods. The expert values of the membership function for qualitative parameters within the criterion F_1 on a subset of the selected alternatives are presented as follows: availability of the developer' expert consultation, respectively {0.9, 0.4, 0.6, 0.2, 0.3}, possibility of adjusting the functioning algorithms {0.8, 0.1, 0.8, 0.1, 0.1}, where the expert values of membership functions are ordered for the above five alternatives. Note that, as mentioned above, the value equal to 1.0 corresponds to the full availability of consultations or changes, and the value 0.0 means the absence of this possibility. Taking into account the remaining expert values and assuming equal weights of all the F_i membership criteria, in accordance with expression (2.3), the following values are obtained for the convolution of the preference relationships (criteria): $\mu_0(1) = 0.85$; $\mu_0(3) = 0.25$; $\mu_0(5) = 0.7$; $\mu_0(9) = 0.15$; $\mu_0(11) = 0.20$.

Thus, in the example considered, the best (optimal) alternative was receiver No. 1, as a result of intersection of fuzzy sets containing estimates of the alternatives according to the selection criteria.

The scalar generalized index I_s of the proximity of a particular GNSS receiver (the selected alternative) to the "ideal receiver", as indicated above, can be represented in the form (2.1), and the weighting factors for the preferences criteria are still selected taking into account (2.2).

In general, the distribution of the criteria weights given by the vector $\boldsymbol{J}$ can be arbitrary within the limits of the indicated constraint (2.2). If it is necessary to narrow the search field, some of the criteria can be assigned a zero weight, which excludes the corresponding parameters from the analysis.

Figure 2.5 shows the results of calculating the scalar generalized indicator I_s for various weight ratios of the preference criteria under consideration. In Fig. 2.5a–e, the weight of one of the criteria is 1; i.e., the remaining criteria are not considered, and in Fig. 2.5f, all the considered preferences criteria F_1–F_5 are taken into account with the same weight $j_i = 0.2$, $i = 1$–5.

The variety of results shown in Fig. 2.5a–e does not allow selection of the optimal alternative for the set of all considered criteria, but allows quick (especially with a large database of analyzed alternatives) selection of the best alternative according to one of the specified criteria.

The analysis of the data shown in Fig. 2.3f shows that the GNSS receiver No. 17 is the optimal alternative with equal importance for all criteria. Its characteristics for the aggregate of all equivalent preference criteria are closest to the "ideal receiver."

Now the question arises: to which extent is the conclusion certain? What happens if we vary the weights of the preference criteria?

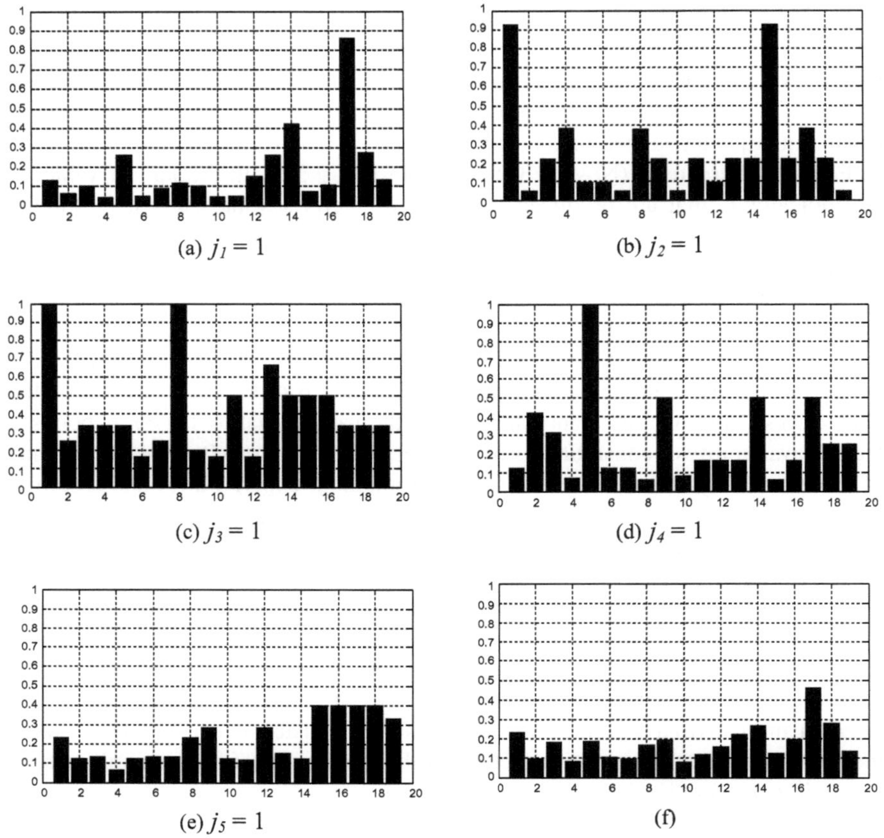

Fig. 2.5 Scalar generalized indicator I_S for various alternatives

To generalize the results of the alternative analysis in the problem of selecting a satellite receiver, we introduce an integrated quality indicator $\tilde{I}_s$ of built-in GNSS receivers that characterize the efficiency of its integration into an SLS or CAS.

The integral quality indicator $\tilde{I}_s$ is defined as follows:

$$\tilde{I}_S = \frac{1}{W} \sum_{(j_i)} \sum_{(k)} I_s(k;\, j_i), \tag{2.5}$$

where the summation is first performed within a particular criterion in accordance with the total number of gradations in the variation of the variable weight of the criterion, and then over all independently variable preference criteria j_i. The multiplier 1/W in (2.5) provides the normalization condition: $\tilde{I}_S \leq 1$.

We note that in the case we are considering, $k \leq 5$, and the weight factors j_i can be selected arbitrarily, taking into account the constraint (2.2).

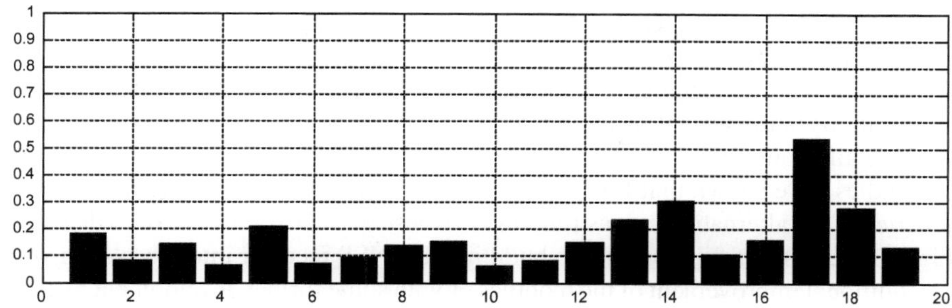

Fig. 2.6 Integral quality indicator $\tilde{I}_s$ with proportional averaging

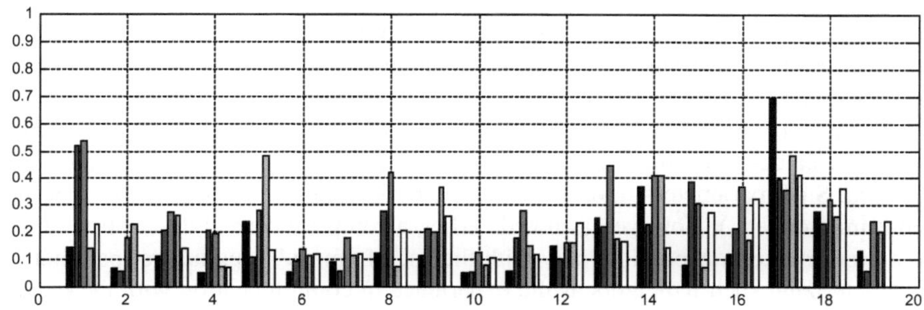

Fig. 2.7 Integral quality indicator $\tilde{I}_s$ with the prevailing criterion

Thus, expression (2.5) represents the average value of the generalized indicators I_s for the corresponding alternatives when varying the weight factors.

Various alternatives corresponding to certain modifications of built-in GNSS receivers will be compared over a wide range of variations in significance coefficients of partial criteria. To obtain an integral generalized indicator, we can vary the weights of the criteria in the entire possible range of variations (from $j_i = 0$, $i = 1, 2, \ldots, 0.5$, when this criterion is of no interest, up to $j_i = 1$, $i = 1, 2, \ldots, 0.5$, when all other criteria are not important). The weights of the remaining criteria will always be selected taking into account the constraint (2.2).

Figure 2.6 shows the results of the analysis of 19 alternatives using the integral quality indicator $\tilde{I}_S$, the value of which is plotted along the ordinate axis. It was assumed that the weight of the criterion F_1 varied in the range from 0.2 to 0.8 with increments of 0.1. Figure 2.7 shows the results obtained for the scalar generalized indicator I_s for the same 19 alternatives represented by sets of particular characteristics, for cases where a weight of 0.8 (the prevailing criterion) was successively assigned for all five criteria F_1–F_5 considered.

The analysis of the calculation results presented in Figs. 2.6 and 2.7, as well as more detailed studies [1–4] of the behavior of the quality indicators I_s and $\tilde{I}_s$, show that the use of the integral quality indicator $\tilde{I}_s$ has some advantage over the generalized indicator I_s. This is due to a decrease in the influence of the subjective factor in the expert assignment of the criteria weights.

At the same time, the use of both indicators is stable (it is invariant in some area of variations in the weight factors of the partial criteria) and allows unique selection of the optimal alternative.

Studies have shown that for the same importance of criteria, the quality characteristics of the alternatives are as follows: The best alternatives are the ones that have high indicators of the possibility to build in the GNSS receiver into the system with no additional improvement of the modules or with some further development by their manufacturer as soon as possible, and low cost, the first aspect being dominant.

It is found that from a set of alternatives comprising 100 elements, the same alternatives are the best when changing significance coefficients of partial criteria in the range from 0.02 (practically not essential) to 0.8 (very important). This allows the construction of strategies for quick searching for the desired options with given weight factors of partial criteria [4].

The results obtained with the above method of searching for the optimal alternative were calculated using "clear" parameters characterizing the selected preference criteria. Such parameters have a specific numerical value (volume, weight, power consumption, etc.). However, as noted earlier, a complete correct description of the preference criteria is possible only with the use of "fuzzy" (qualitative) parameters characterizing such indicators as the experience in communication with the manufacturer, consumer feedback on a particular product, etc. Methods of the "fuzzy" sets theory are now widely used in the analysis of multiple factor processes in chemistry, biology, and intelligent decision-making systems [1, 6, 9–13]. The methods of the "fuzzy" sets theory can be used on the basis of the methodological approaches presented in [4]. In this case, instead of the scalar generalized indicator I_s, it is possible to use the integral quality indicator $\tilde{I}_s$, which will additionally reduce the influence of subjective factors in the formation of expert estimates.

Let us generalize the studies carried out earlier on optimizing the process of selecting a satellite receiver for the SLS and CAS.

In general, the partial criteria, the way of combining partial criteria into an integral quality indicator, a system of constraints and individual variables can be described as fuzzy.

In addition, there are situations where constraints cannot be clearly rationalized.

For example, even such a parameter as the operating temperature range for many alternatives can be considered as fuzzy, since the indicators specified by the manufacturer often differ from those implemented in products, especially for new types of GNSS receivers.

Then some parameters, which were considered as constraints, should be translated into the category of fuzzy parameters [13].

In the general form, the problem of selecting the optimal alternative is formulated as follows. Let X be a set of alternatives (modifications of satellite navigation receivers); x be the elements of this set ($x \in X$); $F_k(x)$ be objective functions corresponding to the criteria for selecting the best alternatives, $k = 1, \ldots, K$; and $C_m(x)$ be constraints $m = 1, \ldots, M$. The objective functions and constraint functions contain qualitative and quantitative descriptions. It is required to find those alternatives that correspond to the objective functions and constraints as much as possible.

To obtain the solution, we use the methods of the fuzzy sets theory [6]. To each criterion and to each constraint, we assign the membership functions $\mu_{F_k}(x), \mu_{C_m}(x)$, and the significance coefficients λ_k and ν_m, whereby for the significance coefficients the following relation shall be satisfied:

$$\sum_{k=1}^{K} \lambda_k + \sum_{m=1}^{M} \nu_m = 1. \tag{2.6}$$

The fuzzy solution D is also a fuzzy set on the set of alternatives X with the membership function $\mu_D(x)$,

$$D = F \cap C, \tag{2.7}$$

$$\mu_D(X) = \mu_F(X) \cap \mu_C(X), \tag{2.8}$$

where F and C are fuzzy sets of targets (preference criteria) and constraints with membership functions $\mu_F(x)$ and $\mu_C(x)$, respectively. The membership function $\mu_D(x)$ is a measure of how much the alternative x satisfies both criteria and constraints at the same time.

Further, on the set of solutions, we need to find an optimal solution D_0, that is, the one that most closely belongs to all the preference criteria of and constraints:

$$D_0 : \underset{(X)}{\mathrm{Max}} \{\mu_D(X)\}. \tag{2.9}$$

The convolution of the partial criteria F_k(x), k = 1, …, K and the constraints $C_m(x)$, m = 1, …, M is performed in accordance with the following formulas:

$$F = \bigcap_{k=1}^{K} \lambda_k F_k, \tag{2.10}$$

$$C = \bigcap_{m=1}^{M} \nu_m C_m, \tag{2.11}$$

taking into account the constraint (2.6).

Operations described by Formulas (2.9–2.11) mean that for each alternative $x \in X$ there is a degree of membership to all its partial criteria and constraints. Moreover, it is essential that the smallest value of the weighted membership function is determinant, that is, no matter how good the indicators for the other criteria and constraints, only the smallest indicator is taken into account in the analysis of the alternative.

Selecting one or more alternatives with the largest value μ_D, according to rule (2.9) we obtain the solution to the problem of selecting the best alternatives. Thus, if the significance coefficients λ_k and ν_m are known, such that condition (2.10) is satisfied, the membership functions of each alternative to all private criteria, $\mu_{F_k}(x)$, $k = 1, \ldots, K$ and all constraints, $\mu_{C_m}(x)$, $m = 1, \ldots, M$ are known, then the algorithm for searching the best alternative is described, in accordance with Formulas (2.3–2.6), by the relation:

$$D_0: \ \underset{X}{\mathrm{Max}}\left\{\mu\left[\bigcap_{k=1}^{K}\lambda_k F_k\right] \cap \mu\left[\bigcap_{m=1}^{M}\nu_m C_m\right]\right\}. \tag{2.12}$$

Now let us find the alternative membership functions to partial criteria and constraints. For some problems, the membership functions $\mu_{F_k}(x)$ and $\mu_{C_m}(x)$ for each alternative $x \in X$ can be specified either on the basis of statistical calculations or by expert evaluation. Then, for known λ_k and ν_m, the problem of selecting the best alternative is solved by the method described above. In the case of selecting a GNSS receiver for the user navigation equipment, it should be borne in mind that each objective function (a partial criterion and a specific constraint) depends on several parameters, both qualitative and quantitative,

$$F_k = F_k\left(t_{k1}, t_{k2}, \ldots, t_{k\Omega(k)}\right) = F_k(t_{k\omega}), \quad \omega \in [1, \ \Omega(k)], \tag{2.13}$$

$$C_m = C_m\left(t_{m1}, t_{m2}, \ldots, \ t_{m\theta(m)}\right) = C_m(t_{m\upsilon}), \quad \upsilon \in [1, \ \theta(m)]. \tag{2.14}$$

The membership functions of the alternative x to the criterion F_k and the constraint C_m are defined as follows:

$$\mu_{F_k}(X) = \bigcap_{\omega=1}^{\Omega(k)} \lambda_{k\omega}\mu_{F_k}(x, \ t_{k\omega}), \tag{2.15}$$

$$\mu_{C_m}(X) = \bigcap_{\vartheta=1}^{\Theta(m)} \nu_{m\vartheta}\mu_{Cm}(x, \ t_{m\theta}), \tag{2.16}$$

where $\lambda_{k\omega}$ is the significance coefficient of the parameter $t_{k\omega}$ in the criterion F_k, $\sum_{\omega=1}^{\Omega(k)} \lambda_{k\omega} = 1$, $\mu_{Fk}(x, \ t_{k\omega})$ is the membership function of the parameter $t_{k\omega}$ to the criterion F_k, $\nu_{m\vartheta}$ is the significance coefficient of the parameter $t_{m\vartheta}$ in the constraint C_m, $\sum_{\vartheta=1}^{\Omega(m)} \nu_{m\vartheta} = 1$, $\mu_{Cm}(x, \ t_{m\vartheta})$ is the membership function of the parameter $t_{m\vartheta}$ to the constraint C_m.

The Formulas (2.13–2.16) include qualitative and quantitative parameters.

The membership functions $\mu_{F_k}(x, \ t)$ are calculated as follows.

For clear (quantitative) parameters $t(x)$, the alternative x can meet the best corresponding characteristic $t(0)$ of the GNSS receiver with the ratio of this parameter value to the corresponding parameter of the ideal receiver $t(0)$; that is, its membership function to a partial criterion is a function of the following form:

$$\mu_{Fk}(x, t_{k\omega}) = \begin{cases} \frac{t_{k\omega}(x)}{t_{k\omega}(0)}, & \text{if } t_{k\omega} < t_{k\omega}(0), \\ 1, & \text{if } t_{k\omega} \geq t_{k\omega}(0), \end{cases} \tag{2.17}$$

For clear parameters defining the constraints C_m, the similar expression is true:

$$\mu_{Cm}(x, t_{mv}) = \begin{cases} \frac{t_{mv}(x)}{t_{mv}(0)}, & \text{if } t_{mv} < t_{mv}(0), \\ 1, & \text{if } t_{mv} \geq t_{mv}(0), \end{cases} \tag{2.18}$$

For fuzzy parameters, either the membership functions of each alternative to the criterion for a particular parameter value (set of parameters) $t_{k\omega}$ $\mu_{Fk}(x,\ t_{k\omega})$ or formula dependencies of the recalculation of the parameters t(x) into the objective membership functions of partial criteria (constraints) should be known.

In the first case, $\mu_{Fk}(x,\ t_{k\omega})$ known for fuzzy parameters and membership functions calculated from Formula (2.17) are substituted into Formula (2.13). The quantity $\Omega(k)$ in Formula (2.13) is the total number of parameters corresponding to the criterion F_k, among which ω_1 are clear (quantitative) and ω_2 are fuzzy (qualitative) parameters, $\omega_1 + \omega_2 = \Omega(k)$.

Similarly, for fuzzy parameters, known $\mu_{Cm}\ (x,\ t_{m\vartheta})$ and membership functions of x to constraints calculated by Formula (2.13) are substituted into Formula (2.18). To describe the constraint C_m, ϑ_1 clear and ϑ_2 fuzzy parameters are used, such that $\vartheta_1 + \vartheta_2 = \Theta(m)$.

In the second case, when the ratios $\varphi_{k\omega}$ of the recalculation of parameters $t_{k\omega}$ into the objective function F_k are known, it is possible to define the membership function of the alternatives to the criterion F_k by setting the membership functions $t_{k\omega}$ for the set of admissible alternatives $\mu_X\ (x,\ t_{k\omega})$.

Using LA. Zadeh's principle of generalization [6], the procedure for fuzzy variable conversion can be written in the form:

$$\mu_{F_k}\ (\mathrm{x},\ t_{k\omega}) = \mu_X(\mathrm{x},\ \varphi_{k\omega}(t_{k\omega})) = \sup_{\mathrm{t}_{k\omega} \in \varphi_{k\omega}^{-1}(t_{k\omega})} \mu_X\ (\mathrm{x},\ t_{k\omega})\,. \tag{2.19}$$

Similarly, the membership function of the alternative $x \in X$ with the parameter $t_{m\vartheta}$ to the constraint C_m with the known relations $\psi_{m\vartheta}$ of the recalculation of parameters $t_{m\vartheta}$ into the function of the constraint $\mathrm{C_m}$ and the known membership function $t_{m\vartheta}$ to the set of admissible alternatives $\mu_x(x,\ t_{m\vartheta})$:

$$\mu_{C_m}\ (\mathrm{x},\ t_{m\vartheta}) = \mu_X(\mathrm{x},\ \psi_{m\vartheta}(t_{m\vartheta})) = \sup_{\mathrm{t}_{m\vartheta} \in \psi_{m\vartheta}^{-1}(t_{m\vartheta})} \mu_X\ (\mathrm{x},\ t_{m\vartheta})\,. \tag{2.20}$$

We consider possible algorithms for convolution of partial criteria and constraints. There are various ways of combining information for multi-criteria optimization on a fuzzy set of alternatives. We indicate some of them based on the use of the intersection of sets of partial solutions and constraints.

Taking into account the notations introduced above, the solution D is written in the form of one of the following sets of formulas:

First set:

$$\mu_F(x) = \text{Min}\left\{\lambda_1\mu_{F_1}(x),\ \lambda_2\mu_{F_2}(x),\ \ldots,\ \lambda_K\mu_{F_K}(x)\right\}, \tag{2.21}$$

$$\mu_C(x) = \text{Min}\left\{\nu_1\mu_{C_1}(x),\ \nu_2\mu_{C_2}(x),\ \ldots,\ \nu_M\mu_{C_M}(x)\right\}, \tag{2.22}$$

$$\begin{aligned}\mu_D(x) = \text{Min}\{&\lambda_1\mu_{F_1}(x),\ \lambda_2\mu_{F_2}(x),\ \ldots,\ \lambda_K\mu_{F_K}(x);\\ &\nu_1\mu_{C_1}(x),\ \nu_2\mu_{C_2}(x),\ \ldots,\ \nu_M\mu_{C_M}(x)\}\end{aligned} \tag{2.23}$$

Second set:

$$\mu_F(x) = \text{Min}\left\{\mu_{F_1}^{\lambda_1}(x),\ \mu_{F_2}^{\lambda_2}(x),\ \ldots,\ \mu_{F_K}^{\lambda_K}(x)\right\}, \tag{2.24}$$

$$\mu_C(x) = \text{Min}\left\{\mu_{C_1}^{\nu_1}(x),\ \mu_{C_2}^{\nu_2}(x),\ \ldots,\ \mu_{C_M}^{\nu_M}(x)\right\}, \tag{2.25}$$

$$\begin{aligned}\mu_D(x) = \text{Min}\Big\{&\mu_{F_1}^{\lambda_1}(x),\ \mu_{F_2}^{\lambda_2}(x),\ \ldots,\ \mu_{F_K}^{\lambda_K}(x);\\ &\mu_{C_1}^{\nu_1}(x),\ \mu_{C_2}^{\nu_2}(x),\ \ldots,\ \mu_{C_M}^{\nu_M}(x)\Big\}\end{aligned} \tag{2.26}$$

Third set:

$$\mu_F(x) = \text{Min}\left\{\frac{\mu_{F_1}(x)}{\lambda_1},\ \frac{\mu_{F_2}(x)}{\lambda_2},\ \ldots,\ \frac{\mu_{F_K}}{\lambda_K}\right\}, \tag{2.27}$$

$$\mu_C(x) = \text{Min}\left\{\frac{\mu_{C_1}(x)}{\nu_1},\ \frac{\mu_{C_2}(x)}{\nu_2},\ \ldots,\ \frac{\mu_{C_M}(x)}{\nu_M}\right\}, \tag{2.28}$$

$$\begin{aligned}\mu_D(x) = \text{Min}\Bigg\{&\frac{\mu_{F_1}(x)}{\lambda_1},\ \frac{\mu_{F_2}(x)}{\lambda_2},\ \ldots,\ \frac{\mu_{F_K}(x)}{\lambda_K};\\ &\frac{\mu_{C_1}(x)}{\nu_1},\ \frac{\mu_{C_2}(x)}{\nu_2},\ \ldots,\ \frac{\mu_{C_M}(x)}{\nu_M}\Bigg\}\end{aligned} \tag{2.29}$$

The optimal solution, as before, is written as:

$$D_0:\ \underset{(X)}{\text{Max}}\left\{\mu_D(X)\right\}. \tag{2.30}$$

With equal partial criteria and constraints, all known methods of combining these yield the same results:

$$\mu_D(x) = \left\{\mu_{F_1}(x),\ \mu_{F_2}(x),\ \ldots,\ \mu_{F_K}(x);\ \ \mu_{C_1}(x), \mu_{C_2}(x),\ \ldots,\ \mu_{C_M}(x)\right\}. \tag{2.31}$$

Another group of criteria is based on the use of a convex combination of components, taking into account their importance:

$$\begin{cases} D = F + C, \\ \mu_D(x) = \mu_F(x) + \mu_C(x), \\ F = \sum\limits_{k=1}^{K} \lambda_k F_k, \\ C = \sum\limits_{m=1}^{M} \nu_m C_m, \\ \sum\limits_{k=1}^{K} \lambda_k + \sum\limits_{m=1}^{M} \nu_m = 1, \end{cases} \tag{2.32}$$

where $\mu_D(x) = \sum_{k=1}^{K} \lambda_k\, \mu_{F_k}(x) + \sum_{m=1}^{M} \nu_m \mu_{C_m}(x)$,

$$\mu_{F_k}(x) = \sum_{\omega=1}^{\Omega(k)} \lambda_{k\omega} \mu_{Fk}(x,\ t_{k\omega}),$$

$$\mu_{C_m}(x) = \sum_{\vartheta=1}^{\Theta(m)} \nu_{m\vartheta} \mu_{C_m}(x,\ t_{m\vartheta}),$$

$$\mu_D(x) = \sum_{k=1}^{K} \sum_{\omega=1}^{\Omega(k)} \lambda_k \lambda_{k\omega} \mu_{F_k}(x,\ t_{k\omega}) + \sum_{m=1}^{M} \sum_{\nu=1}^{\Theta(m)} \nu_m \nu_{m\vartheta}\, \mu_{C_m}(x,\ t_{m\vartheta}).$$

The optimal solution, as before, is written as (2.30).

Expressions (2.32) resemble optimized functionals in multi-criteria optimization in the case of clear quantitative characteristics.

As noted above, there is a correlation relationship between the partial preference criteria considered. Above are formulas for combining independent partial criteria into a generalized quality criterion. However, the condition of independence is not always fulfilled. One possible method for taking into account the correlation between the criteria is as follows. The set of alternatives is expanded in accordance with the existing correlation dependence. For example, the criteria $\boldsymbol{F}_1$ (integrability into the FNC (flight navigation complex) structure) and $\boldsymbol{F}_4$ (cost) considered above (see details in [2]) are dependent. The parameter t_2, the possibility of adjusting the functioning algorithms, is included as a component in both criteria. By assigning various values to the parameter t_2 for a particular alternative, we obtain the corresponding values of the membership functions of this alternative to the criteria $\boldsymbol{F}_1$ and $\boldsymbol{F}_4$.

Let t_2 take one of the q possible values, $t_2 \in [t2_1, t2_2, \ldots, t2_q]$. Then, if all m modifications of the GNSS receiver are available in the market, we should analyze, instead of one particular alternative, its q varieties, with each of the q values of the parameter t_2 having pairs of values of the objective functions $\{\boldsymbol{F}_1, \boldsymbol{F}_4\}$.

Thus, using the fuzzy sets theory, the main methodological approaches to the selection of the GNSS receiver for the SLS and CAS are formulated. A set of alternatives is determined and they are structured by the method of expert evaluation. The optimization method is illustrated with a reduced set of criteria and constraints. The results are presented in the form of a table of the membership functions of the alternatives to each criterion, indicating the degree of their proximity to the optimal solution.

With the proposed methodological approaches, it is possible to select such a GNSS receiver that will provide the optimal solution of problems in terms of flight safety provision.

This approach makes it possible to trace the dynamics of the optimization process and to identify groups of factors that have a dominant influence on the decision-making process. The method makes it possible to increase the volume of affecting factors, as well as to take into account the correlation relationships between individual parameters and criteria.

Various options of information convolution using clear and fuzzy approaches, both to the preference criteria and to the alternatives parameters, give practically the same results when the preference criteria weights are varied over a wide range.

The methodological approach proposed above for the SLS and CAS development makes it possible to address the selection of the main functional core of these systems uniformly, significantly reduces the time needed to find the required solution, and, on the whole, accelerates the process of developing and commissioning the equipment.

We now turn to the analysis of the basic SLS and CA functions and structures in order to determine ways to enhance these systems for improving flight efficiency and safety.

2.2 Methods for Building the Structure of the Ground and Onboard Radioelectronic Complexes of Satellite-Based Landing Systems with Augmentations of Global Navigation Satellite Systems

In order to determine the most effective ways of using the SLS to enhance flight safety, it is advisable to consider their functions and structure, as well as to identify those elements the improvement of which is most expedient. The most general form of the SLS structural diagram is shown in Fig. 2.8. The SLS structure includes LAAS located in the landing aerodrome, and the onboard GNSS/LAAS equipment.

Requirements for the LAAS technical characteristics were considered in Sect. 1.4 of the present paper and are given in documents [14–16]. The LAAS shall perform the

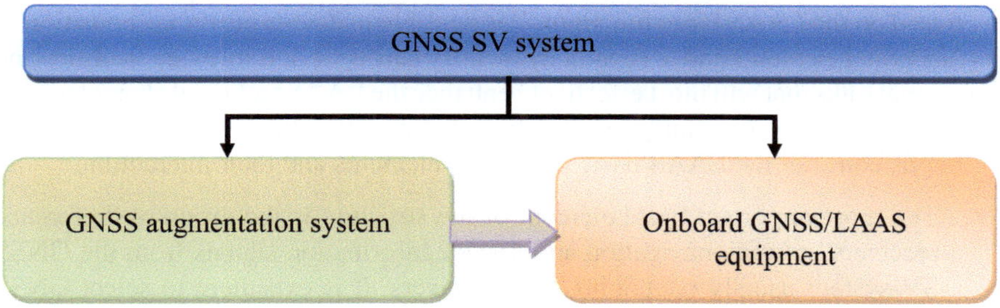

Fig. 2.8 General SLS structure diagram

function of supporting precise approaches and categorized landings to aerodromes and sites, as well as to provide support of other vehicles and procedures that require precise positioning, provided they are located within the LAAS coverage area. The LAAS functional capabilities and characteristics for en route flights, terminal area flights, approaches, and landings depend on the LAAS category characterized by indicators (GAD). At present, three indicators are applied: A, B, or C. The LAAS, to which the minimum requirements are applied, have an indicator "A", and the LAAS satisfying the highest requirements, including those to ensure Category II and III landings has an indicator "C".

Ground-based GNSS augmentation system (LAAS). The main functions of the LAAS, which is one of the elements of the SLS general structure, are [14]:

1. Generation of differential corrections (DCs) for pseudoranges and associated parameters for all satellites in sight of the ground antenna feed system of ground GNSS reference receivers;
   - parameters of the state and characteristics of the LAAS itself;
   - parameters for the final approach segment;
   - parameters of predicted availability of the range sources.
2. Transmission of generated messages over the radiochannel.
3. Integrity monitoring for:
   - data of observable navigation satellites;
   - generated differential data;
   - the radiochannel and the messages transmitted over it.
4. Monitoring the continuity of generated and transmitted data;
5. Monitoring own performance;
6. Control of the equipment operation and LAAS operation modes;
7. Registration of the LAAS messages, parameters, and operating modes transmitted over the radiochannel, faults, failures, and other operational disturbances, external control actions, and environmental conditions;
8. Receiving, storage, update, and transmission of auxiliary, service and other information over the LAAS—remote control center communication line.

Based on the LAAS basic functions, it is possible to determine the composition of its functional elements and to separately identify hardware and software facilities and functional links that will most effectively enhance the LAAS technical characteristics (accuracy, integrity, continuity, availability) that directly affect flight safety.

Let us consider the LAAS basic functional elements and their interaction.

1. The most important basic element of the structure is the group of "reference receivers" of radionavigation and service information signals from the GNSS SV system usually containing four receivers. It is expedient to select GNSS receivers for this LAAS functional element according to the procedure described in Sect. 2.1 of this paper. It is advisable to add an SBAS signal receiver to this LAAS structure element, which can significantly improve the LAAS technical characteristics as a whole, since there is another additional channel of information on the GNSS integrity.
2. The computing device (with interface modules) for calculation and formation of DCs and other information related to DCs. This device is the hub for the entire information circulating in the LAAS and provides software and algorithmic implementation of new ways to improve the LAAS technical characteristics. Usually, this computing device is reserved.
3. The reference receiver for monitoring the GNSS integrity is the device that allows the exclusion of the navigation signals that are not correctly formed on the SVs from the navigation data flow. This is especially important, because if this signal is used to calculate the DCs, then their use can lead to distortion of the end user data and loss of the LAAS integrity.
4. The device for monitoring the integrity of SV data and the calculated DCs. In this device, new ways of improving the integrity of the LAAS data can be implemented.
5. The device for receiving, storing, adjusting, and using the database necessary to ensure the DC generation. In this device, it is also proposed to implement new methods for increasing the accuracy and integrity of the LAAS data.
6. The device for transmission of DCs, glide path parameters, and LAAS data for users over the VHF radiochannel. Pseudosatellites belonging to the LAAS and operating both in the GNSS frequency range and at other radiofrequencies can also be included here.
7. The device for monitoring the integrity of data transmitted over the radiochannel.
8. The device for automatic control of the LAAS operation modes, as well as for control via external commands.
9. The device for registering performance parameters, output data, and other information.
10. Means of ensuring the LAAS operation (power sources, protection from external electromagnetic radiation, etc.).

Figure 2.9 shows the generalized LAAS diagram, which includes the listed functional elements and the relationships between them.

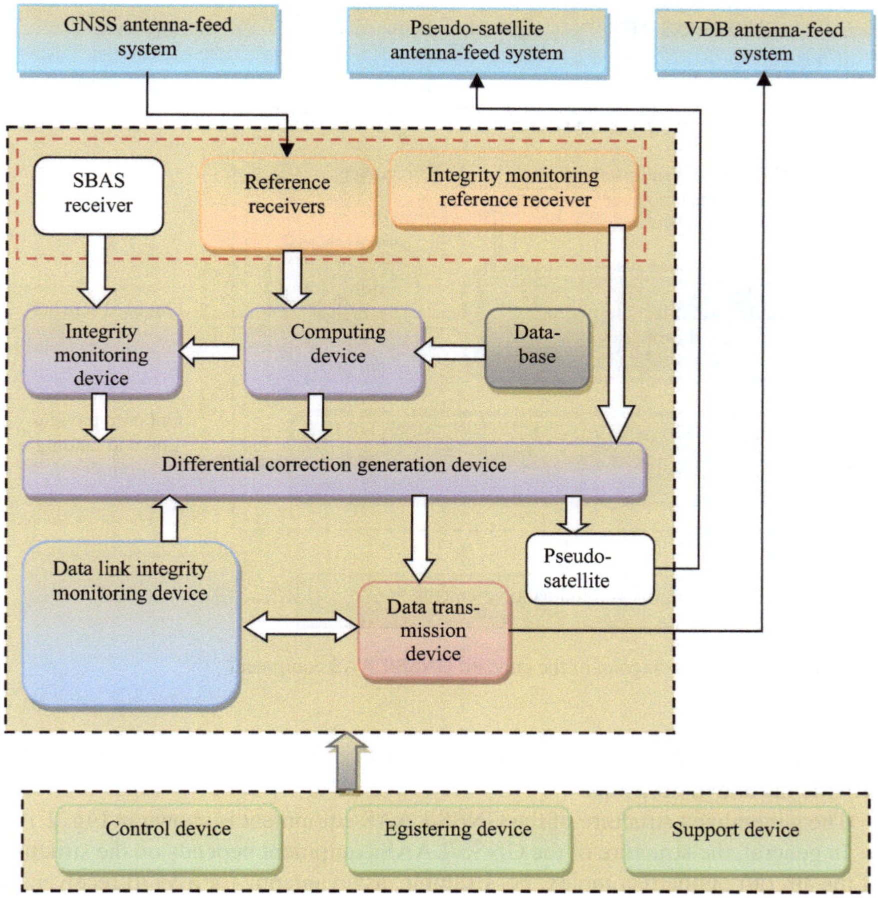

Fig. 2.9 Generalized LAAS diagram

The feature of the diagram shown in Fig. 2.9 is the SBAS receiver and pseudosatellites included in its composition, the use of which allows improving the LAAS performance.

Functions of the SLS onboard subsystem (GNSS/LAAS equipment). The main functions of the GNSS/LAAS equipment are

- reception of differential data and data on the final approach segment (FAS)—VDB receiver functions;
- navigation and positioning (PAN) provision;
- generation of signals of deviation from the calculated approach path;
- generation, output, and indication of navigation and landing information for users;
- control of own performance with the output of a quality indicator of the information provided.

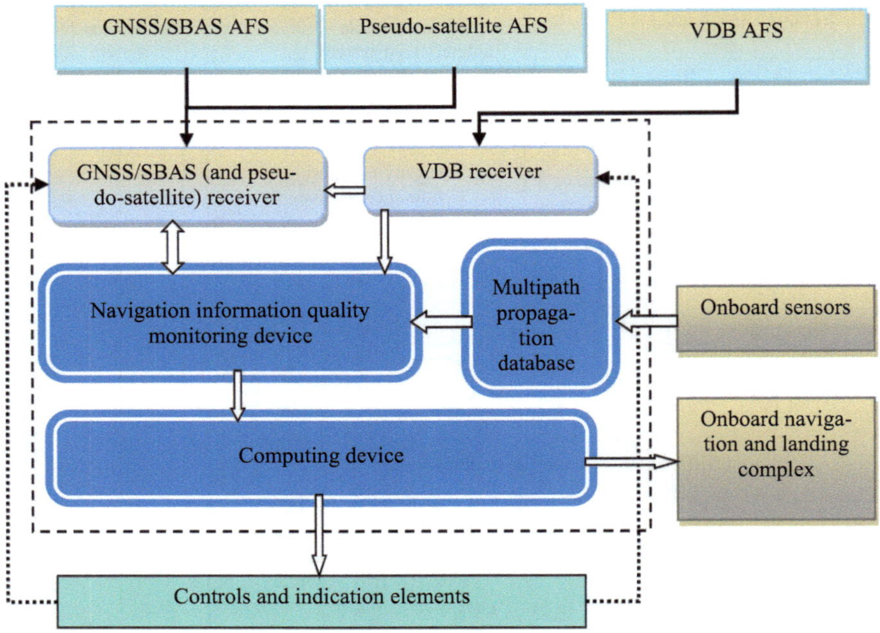

Fig. 2.10 Generalized diagram of the onboard GNSS/LAAS equipment

The onboard GNSS/LAAS equipment includes all hardware and software that provide the above functions.

The generalized structure of the GNSS/LAAS equipment is shown in Fig. 2.10.

In general, the structure of the GNSS/LAAS equipment depends on the structure of the aircraft onboard complex. For example, as an antenna for a VDB receiver, an ILS localizer antenna can be used, and the control panel of the flight management system can be used as controls and indication elements.

The principal feature of the structure presented in Fig. 2.10 is the presence of a quality monitoring device for navigation information and a database of the aircraft performance as an object of radiowaves reflection causing multipath propagation of signals from the SV to the onboard antenna of the GNSS receiver.

The onboard GNSS/LAAS equipment provides data on position, velocity and time (PVT).

If the equipment uses differential corrections from the LAAS, then the output PVT data of the onboard GNSS/LAAS equipment shall meet the requirements in [17].

If the GNSS/LAAS equipment does not use differential corrections from the LAAS, then the output PVT data shall meet the general requirements for the GNSS equipment.

The functionality and characteristics of the onboard GNSS/LAAS equipment depend on its class (A or B).

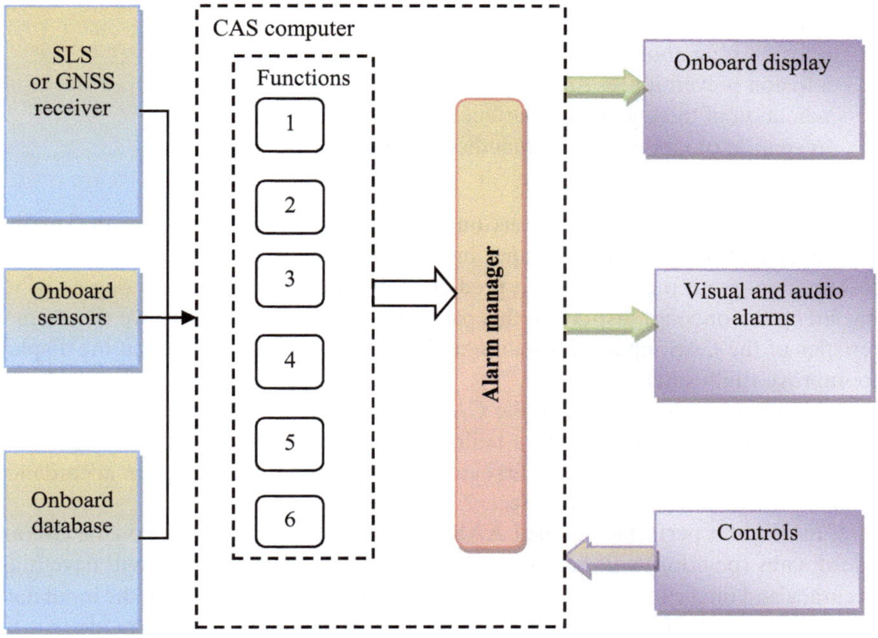

Fig. 2.11 Generalized diagram of the collision avoidance system construction

Class A characterizes onboard equipment with minimum requirements for RMS errors in the determination of the corrected pseudorange, and Class B characterizes high-precision equipment designed to support ICAO Category II and III landings.

Consideration of the above structural diagrams makes it possible to determine those functional elements, the development of which will have the greatest effect in terms of improving the SLS technical characteristics as a whole.

Such elements include LAAS integrity control devices (see Fig. 2.9) and a device for monitoring the quality of navigation information in the GNSS/LAAS equipment (see Fig. 2.10).

Next, let us briefly consider the main elements that determine the CAS structure.

2.3 Methods and Rules for the Development of a Collision Avoidance System with the Use of Global Navigation Satellite System Technologies

Figure 2.11 shows the generalized structural diagram of the CAS construction and interoperable onboard sensors and devices from the aircraft navigation system.

The main CAS functions (see Sect. 1.4) are:

1. collision avoidance (basic modes);
2. early warning about the ground proximity;
3. collision prevention;
4. indication of the underlying surface;
5. prevention of landing on an unauthorized runway;
6. runway movement monitoring.

The effect of all these functions on flight safety is determined by the quality (accuracy and integrity) of the input information, as well as the representation of information about the current and predicted aircraft position and the degree of its hazard on the onboard display. In this paper, it is shown *how* to improve the characteristics of the CAS input information and *what* should be indicated on the display to improve flight safety.

To implement functions 2–6, the CAS uses navigation data from the GNSS. In this case, the CAS may include a built-in (integrated) GNSS receiver or use data from an external receiver. In the first case, it is expedient to select it in accordance with the methodology set out in Sect. 2.1 of this paper.

If the flight is performed in the LAAS coverage area, the input data on the aircraft coordinates (position) will be obtained in the differential mode, i.e., will have high accuracy and integrity, and for flights outside the LAAS coverage area, the input data will correspond to the characteristics of the GNSS standard operation mode.

The composition of the onboard sensors with which the CAS interacts (see Fig. 2.11) includes, as a minimum: a radio altimeter, an air data system, and an instrument landing system.

In addition, an inertial system, an onboard radar and other sensors of navigation information can be used.

All CAS functions are implemented using a digital computing device (computer), which includes all the necessary interface modules that ensure the information exchange between the CAS components.

The onboard CAS database includes the following mandatory elements: a digital terrain model, a database of airports, and a database of artificial obstacles. The same base includes the database of aircraft performance used in the formation of protective spaces for the implementation of early warning and collision avoidance functions, etc.

The alarm manager (see Fig. 2.11) generates and outputs the most priority alarm (warning) from all alarms received by the current time from all functions implemented in the CAS to the pilot. This ensures unambiguous actions of the pilot in the event of multiple warnings.

It should be noted that the priority of messages in the CAS is rigidly set by regulatory requirements [18].

Considering the fact that about 65% of accidents occur due to piloting errors (see Sect. 1.6), special attention should be paid to the development of some indication for the pilot, which will prevent the aircraft from being involved in emergency situations.

The next stage is the development of functional modules that ensure the performance of vertical maneuvers, determination of the turn direction and determination of the possibility to safely continue the flight in a given direction, the use of which during flights under the conditions of hazardous terrain or reduced visibility will also prevent piloting errors.

An important task is the integrated use of the SLS and CAS information, which not only increases the efficiency of the CAS use, but also makes it possible to increase the accuracy and reliability of each of the systems used.

Next, consider the directions and ways for SLS and CAS improvement.

2.4 Directions and Methods to Enhance Satellite-Based Landing Systems and Collision Avoidance Systems

First consider SLSs. In accordance with the generalized SLS diagram, shown in Fig. 2.8, its technical characteristics are determined by the features of the ground-based GNSS augmentation system (LAAS) and the characteristics of the onboard GNSS/LAAS equipment.

The main ways to improve SLSs can be determined by analyzing the effect of various factors on the characteristics of the ground and onboard SLS subsystems using GNSS technology.

In the standard operating conditions of the GNSS SVs, the LAAS characteristics depend on the following:

- capability of the LAAS software and algorithmic support to correctly generate DCs and evaluate their quality, as well as to meet data integrity requirements,
- organizational, technical, and program algorithmic measures to eliminate the effect of re-reflections (multipath propagation) on the calculated DCs,
- presence of sources of electromagnetic interference near installation sites of LAAS reference receiver antennas,
- technical characteristics of the LAAS reference receivers,
- technical characteristics of the VDB transmitter and its AFS.

Under the normal operating conditions of the GNSS SVs, the characteristics of the onboard GNSS/LAAS equipment depend on the following:

- characteristics of software and algorithmic support of the onboard equipment regarding generation of estimates for the current aircraft position;
- level of the re-reflected SV signals from the aircraft structure elements;
- presence of sources of electromagnetic interference near installation sites of the onboard antenna;
- technical characteristics of the onboard GNSS receiver;
- technical characteristics of the VDB receiver and its AFS.

Let us analyze these factors. The LAAS deployment implies special requirements when selecting installation sites for reference receivers (RRs) antennas and the VDB transmitter antennas. When planning antenna installation sites, requirements for minimum restrictions on cutoff angles should be provided. The installation site of the

LAAS RR is selected in an area free from obstacles that interfere with the reception of satellite signals at the lowest possible elevation angles. In general, any masking of GNSS satellites at elevation angles above 5° will lead to a deterioration of the system availability. The design and location of LAAS RR antennas shall limit the multipath effect, which interferes with a desired signal. The installation of antennas near the earth's surface reduces the multipath caused by reflections under the antenna. The height of the installation is chosen to be sufficient to prevent the antenna from being covered with snow or interference from the maintenance personnel or surface vehicles. The antenna should be located in such a way that any metal structures, such as fans, pipes and other antennas, are outside its near field. The RR antennas are located in such places that the conditions of multipath propagation of radiowaves for various antennas are different and that the traffic movement does not introduce additional reflections.

The VDB transmitter antenna on the LAAS is located in such a way that there is a line of sight from the antenna to any point within the coverage area for all possible approach paths.

Generally speaking, increasing the height of the VDB transmitter antenna may be necessary to provide a proper signal power level for the aircraft at low altitudes, but it can also lead to unacceptable gaps in the radiation pattern due to the multipath propagation of the radiowaves in the desired coverage area. The optimum antenna height should be selected on the basis of analysis and taking into account guaranteed compliance with the requirements for the signal power level over the LAAS entire coverage area. It is also necessary to take into account the effect of the earth's surface, buildings, and structures on the multipath effect.

The most important LAAS function is monitoring errors in the ephemeris and GNSS failures through a number of methods. These include:

(a) Increase the RR antenna diversity on the LAAS. Greater diversity helps improve the detection of the minimum detectable error (MDE).
(b) The use of data from the space-based GNSS augmentation system (SBAS). Since the SBAS provides monitoring of satellite characteristics, including ephemeris data, the integrity information transmitted by the SBAS can be used as an indication of the ephemeris validity. The ground-based SBAS subsystem uses GNSS receivers installed with large spatial separation and thus provides optimal ephemeris monitoring characteristics. As a result, it becomes possible to identify small MDE values.
(c) Ephemeris data monitoring. This method involves comparison of the ephemeris transmitted by SVs during the successive passage of satellites over the LAAS. In this case, it is assumed that the only reason for the fault is the error in the ephemeris transmitted by the network of ground control and GNSS information loading stations. In order for this method to ensure the required integrity, it is necessary to exclude the possibility of failures due to unauthorized maneuvers of satellites.

At present, more attention is paid to monitoring systems. The characteristics of the monitoring device (for example, MDEs that it detects) shall be based on the requirements for the integrity loss risk and the failure model, the protection from which shall be provided by this monitoring device. The rate for failures of the GPS ephemeris information can be determined on the basis of the requirements defined in [14, 15].

Since the ephemeris error can lead to serious problems in the navigation support, the above GLONASS and GPS characteristics shall be taken into account in the integrity monitoring algorithms.

A standard LAAS processes measurements obtained from 2 to 4 RRs installed in the immediate vicinity of the LAAS reference point. The onboard GNSS/LAAS receiver is protected from large errors or faults in one of the LAAS RR by calculating and applying the integrity parameters B_i transmitted by the LAAS to the aircraft through the VDB channel.

The most important LAAS function is integrity monitoring. The integrity monitoring philosophy includes the following stages (without taking into account the ground-to-air data channel):

- testing navigation satellites in view and selection of ranging sources for the DC generation (the first stage of monitoring);
- testing generated DCs with an iterative procedure for specifying the composition of the ranging sources (the second stage of monitoring).

The third stage of the integrity monitoring is performed in the onboard GNSS/LAAS equipment. The defect detection logic includes:

- a set of algorithms for comprehensive monitoring with a variety of signal tests;
- a set of algorithms for isolating defective ranging sources;
- algorithms of source restoration after elimination of the defect.

At the first stage of testing, three types of integrity monitoring algorithms can be used.

1. Analysis of radiosignal parameters in the receiver (signal level, code structure, etc.). These tests are commonly called SQM (signal quality monitoring) [15]. For integral analysis of the radiosignal quality by the correlation function form, it is desirable to have a special quality receiver (SQR).
2. Analysis of navigation information reliability (satellites "health", correspondence of ephemeris to the almanac, etc.). These tests are called DQM (data quality monitoring).
3. Analysis of the input filter data and the quality of its operation (MQM—measurement quality monitoring).

In the second stage of testing, the following procedures are involved: sigma monitoring to monitor the accuracy of the DC generation and the output DC control based on the use of several reference receivers with generation of B-values (MRCC—multiple reference consistency check and verification of the output data boundaries).

In accordance with the above philosophy of integrity monitoring, algorithms for processing signals and data in the LAAS are shown in Fig. 2.12.

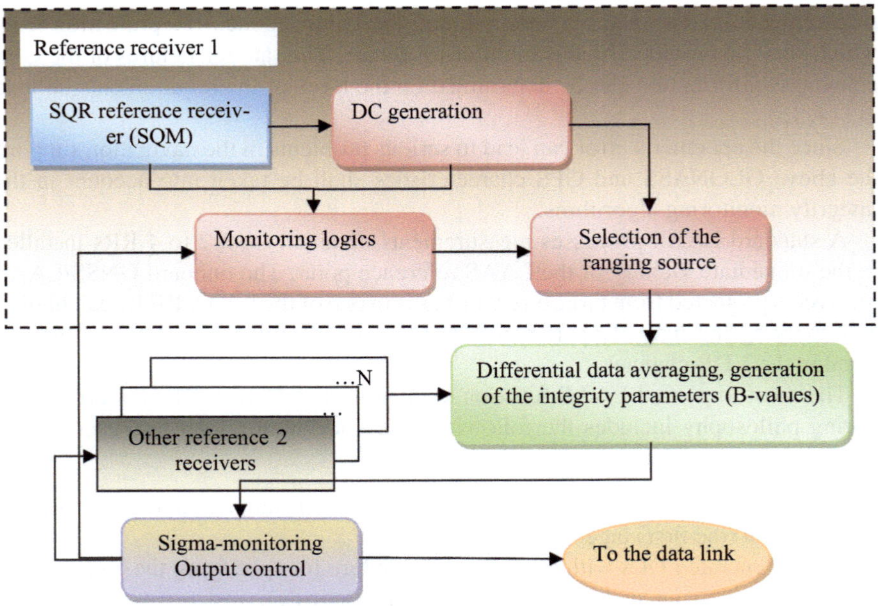

Fig. 2.12 LAAS signals and data processing logics

Experimental studies show that the main contribution to the error in the DC generation is made by effects associated with radiosignal re-reflections [17–21]. Therefore, *all* possible methods against these effects must be provided in the LCMS equipment. Such methods include pre-correlation and post-correlation methods. Pre-correlation methods include spatial (including polarization), time and spectral processing of signals. The most effective is the spatial processing of signals that includes radio-diversity techniques by means of specialized, reflection-resistant antenna systems. Time and spectral processing of signals are implemented in navigation receivers. The simplest time processing to provide protection against re-reflections is to use small time diversity of reference signals in the GNSS receiver correlator (0.1 chip instead of the commonly used 1.0 chip). Post-correlation methods are implemented at the output of GNSS navigation receivers and are part of the integrity monitoring algorithms.

When the GNSS is used, a great role is played by the redundancy of navigation information and a good geometry. Therefore, it is desirable that the LAAS includes pseudosatellites to provide the required characteristics (see Fig. 2.9), which positively influence the accuracy of determining the navigation parameters, especially altitudes, and the availability of measurements.

Thus, the determining factors for ensuring the required technical characteristics of the LAAS are integrity monitoring, protection against re-reflections of navigation radiosignals, and use of pseudosatellites.

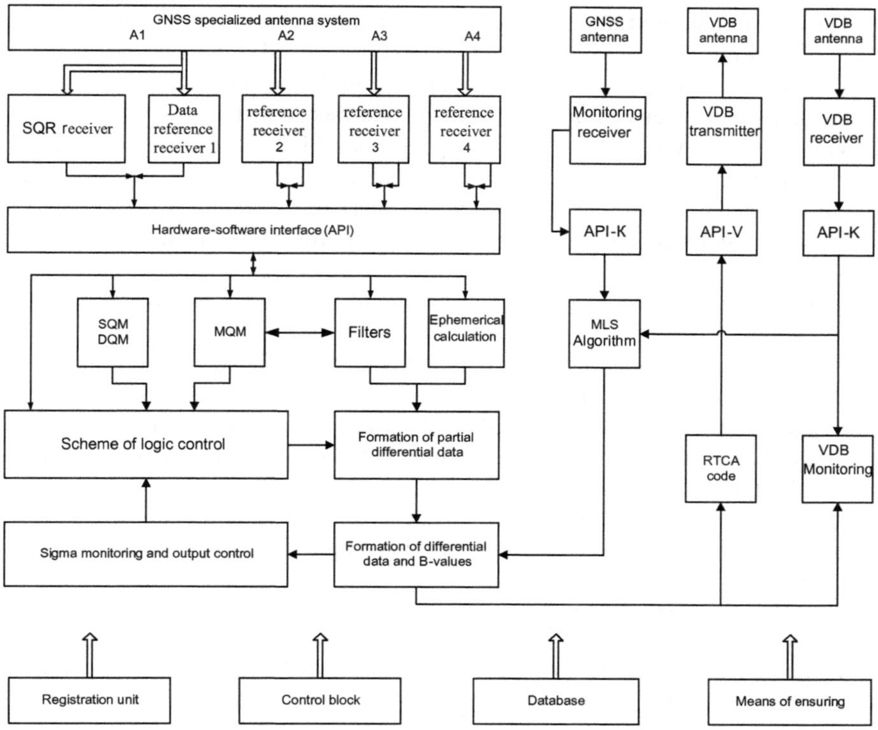

Fig. 2.13 Functional LAAS diagram with elements that enhance flight safety

The implementation of the above integrity monitoring methods determines the functional diagram for LAAS constructing shown in Fig. 2.13. The diagram includes the hardware and software parts separated by the hardware/software interface (HSI).

The navigation data of each reference receiver is used to generate partial differential data: PRC_{ij} (correction), RRC_{ij} (correction rate), and σ_{ij} (error in the correction estimate), where i is the number of the ranging source (satellite), j is the number of the LAAS reference receiver. This process is accompanied by the selection of partial DCs in accordance with the philosophy described above. The DC averaging and the generation of PRC_i, RRC_i, σ_I and B_j follow the majority principle.

The diagram in Fig. 2.13 contains a receiver for the DL signals to monitor its functioning (the transmitted and received DL digital data are compared); it includes integral monitoring of the LAAS operation by means of a separate monitoring receiver that solves the navigation task (the LSM block) using DCs. Note that with separate monitoring for DC and LD generator, such integrated monitoring can additionally improve the reliability of the o ground subsystem operation as a whole. The functional diagram also includes a pseudosatellite capable of providing a specified continuity and availability risk level.

The determination of the LAAS contribution to the corrected pseudorange error (σ_{pr_gnd}) is a rather intricate problem that is not regulated in the current standards [17]. Sources contributing to this error include multipath propagation of satellite radiosignals and receiver noises. In view of the fact that the SLS is designed to provide precise landings per ICAO Category I and higher categories in future, in the construction of the "ground-air" differential data transmission channel, a great attention is paid to interference protection and interference immunity of this channel.

Let us now consider the main ways and directions for improving the onboard SLS subsystem.

In the GNSS/LAAS equipment placed onboard aircraft, the following additional errors occur.

First, it is the contribution of the onboard receiver to the corrected pseudorange error. The maximum value of this contribution can be estimated under the assumption that $\sigma_{receiver}$ is equal to RMS_{pr_air} for the onboard equipment (GNSS/LAAS) with an accuracy of class A. For more precise equipment (class B), more careful calculations are required that are not currently standardized.

Secondly, these are multipath errors due to the influence of the aircraft hull. Multipath errors due to reflections from other objects are usually not taken into account.

However, if the experience of using the LAAS at a particular aerodrome shows that these errors cannot be neglected, they are taken into account by increasing the values of the parameters transmitted by the LAAS, for example, σ_{pr_gnd}. Such conditions usually arise when the aerodromes have hazardous (mountainous) terrain relief or in the presence of a significant number of artificial structures in the vicinity.

Finally, a significant contribution to the errors in the position determination is made by the uncertainty of errors in the ephemeris. Pseudorange errors due to errors in the ephemeris (defined as discrepancies in the true and calculated position of the satellite) are partially decorrelated and, therefore, will be different for receivers onboard the aircraft and in the LAAS.

If users are relatively close to the LAAS reference point, the residual differential error due to errors in the ephemeris will be valid for correcting coarse measurements and calculating the protection levels.

One of the methods for ensuring continuity requirements assumes that there are backup facilities onboard the aircraft, for example, ABAS on the basis of inertial systems and that ABAS provides sufficient accuracy for performing a particular operation.

Thus, the accuracy, integrity, and continuity of the navigation and landing information received onboard the aircraft when using the SLS depends significantly on the presence and nature of re-reflected radiosignals in the area of the LAAS reference receiver antenna installation site, as well as in the area of the GNSS antennas, on the presence of radiointerference in the GNSS operating range and the visibility of navigation SVs in the area of the landing aerodrome.

The development of methods that improve the SLS characteristics by compensating for the effects of these impacts both on the LAAS and onboard the aircraft is one of the goals of this paper.

Now consider CASs. Let us analyze the basic CAS functions and its structure (see Fig. 2.11), which are briefly described above in Sects. 1.4 and 2.3.

Of the basic CAS modes under our consideration, the most interesting is the mode of warning of the excessive deviation from the approach glide path. This mode ensures flight safety during approaches and landings, i.e., at high-risk phases. In conventional CASs, this mode only functions during landings on aerodromes equipped with ILSs. As is known, in Russia such systems are installed in 100 aerodromes maximum out of almost 2500 in operation. Therefore, the use of a synthetic glide path, which can be built on the basis of the GNSS information and aeronautical information from the landing airport, will make it possible to monitor the approach with a synthetic glide path at any aerodrome.

The early ground proximity warning function can additionally improve flight safety with more accurate and reliable GNSS data from the GNSS/LAAS subsystem due to a more accurate determination of the distance to the hazardous terrain and obstacles.

In order to prevent aircraft from encountering an emergency situation during flights under the conditions of hazardous terrain, it is advisable to develop a new function, that is, the function of notifying about the possible collision.

For flight safety improvement, the ergonomics of the information displaying is important regarding the underlying surface presented on the display, since this picture directly affects the "flight image" used by the aircraft crew during piloting.

A separate group of methods to improve flight safety is associated with the movement on the runway after landing in order to prevent the aircraft from overrunning the runway.

Thus, the GNSS technology allows for the modernization of old functions and the introduction of new ones into the CAS in order to improve flight safety.

In the following sections, specific methods and devices will be considered that enhance flight safety when SLSs or CASs are used.

2.5 Conclusions

In this chapter, a methodological approach is presented to the selection of the main functional elements of SLS and CAS radiotechnical complexes based on the multiple factor analysis and fuzzy sets theory, which allows minimizing the influence of subjective factors and shortening the time for searching the optimal alternative. A feature of the proposed approach is the use of elements of an expert navigation-oriented system. This makes it possible to trace the dynamics of the optimization process, as well as to identify groups of factors that have a dominant influence on the decision-making when selecting the basic elements for SLS and CAS construction. The mathematical apparatus that uses this approach is implemented in the MATLAB programming environment. The use of performance databases for the GNSS modules of various manufacturers makes it possible to experimentally test the effectiveness of the proposed methodological approach.

The key SLS and CAS elements affecting their effectiveness are ground and onboard devices for monitoring the quality and integrity of the GNSS information for the SLS; and the functions of displaying the underlying surface, preventing collisions, preventing landings on an unauthorized runway, preventing the aircraft from overrunning the runway for the CAS.

The integrated use of the SLS and CAS allows expanding the area of the SLS use for those aerodromes that are not equipped with ILS-like systems and for those aircraft that do not have radio altimeters (almost all small aircraft). The use of SLS information for the implementation of some CAS operation modes leads to an increase in the accuracy, integrity, and continuity of navigation information.

The SLS analysis shows that the most important areas for enhancing the SLS are monitoring the navigation data integrity, the use of methods for protecting navigation radiosignals from multipath propagation (re-reflections), the use of methods to protect navigation radiosignals from electromagnetic interference, the use of methods to improve the navigation data accuracy; the use of pseudosatellites in the structure of the ground systems of the GNSS augmentations.

The CAS analysis shows that the main area for enhancing the CAS is the development of methods and devices that prevent aircraft from encountering the conditions that cause emergency or warning alarms. The most promising areas are increasing the ergonomics of the displaying the underlying surface, preventing encounters with hazardous situations by taking into account the dynamic characteristics of the aircraft and the nature of the underlying surface, preventing landings on an unauthorized runway, and preventing overrunning the runway.

The analysis of the SLS and CAS construction shows that they are based on GNSS technologies. This is a prerequisite for their integrated use in order to expand the area of use for each of the systems and to improve their efficiency.

References

1. Balyasnikov BN, Kuklev EA, Olyanyuk PV, Sauta OI, Shchennikov DL (2012) Selection of a GNSS receiver for navigation and landing complexes using integrated quality indicators (in Russian)
2. Baburov VI, Ivantsevich NV, Sauta OI (2010) Preference criteria in the problem of selecting a satellite receiver for navigation and landing complexes. In: Peshekhonov VG (ed) XVII St. Petersburg international conference on integrated navigation systems. State Scientific Center of the Russian Federation, JSC Concern CRI Elektropribor pp 425–427 (in Russian)
3. Baburov VI, Ivantsevich NV, Sauta OI (2011) Integral quality indicators for GNSS receivers for navigation and landing complexes. In: Peshekhonov VG (ed) XVIII St. Petersburg international conference on integrated navigation systems. State Scientific Center of the Russian Federation, JSC Concern CRI Elektropribor, pp 327–329 (in Russian)
4. Sauta OI (2012) Methodological approach to the development of radio technical complexes (in Russian)
5. Bellman R, Zadeh L (1976) Decision-making under vague conditions. In: Analysis issues and decision-making procedures. Mir, Moscow, pp 172–215 (in Russian)
6. Stulov AV (2003) Operation of the satellite navigation equipment in aviation. Air transport, Moscow, 326 p (in Russian)

7. Solovyev YuA (2003) Satellite navigation and its applications. Eco-Trends, Moscow, 326 p (in Russian)
8. Orlovsky SA (1981) Problems of decision-making with fuzzy information. Nauka, Moscow, 206 p (in Russian)
9. Meshalkin VP (1995) Expert systems in chemical technology. Chemistry, Moscow, 368 p (in Russian)
10. Yager RR (ed) (1986) Fuzzy sets and theory of possibilities. In: Recent achievements. Radio and Communication, Moscow, 408 p (in Russian)
11. Borisov AN, Alekseev AV, Merkuryeva GV et al (1989) Fuzzy information processing in decision-making systems. Radio and Communication, Moscow, 304 p (in Russian)
12. Borisov AN, Alekseev AV, Merkuryeva GV et al (1989) Fuzzy information processing in decision-making systems. Radio and Communication, Moscow, 304 p (in Russian)
13. Sauta OI (2011) Use of satellite navigation information to improve the efficiency of weapons. In: New technologies. Proceedings of the VIII All-Russian Conference. RAS, Mowcow, pp 106–115 (in Russian)
14. Annex 10 to the Convention on International Civil Aviation. Aeronautical Telecommunications. In: Radio Navigation Aids, 6th edn, vol 1 (in Russian)
15. RTCA DO-246C, GNSS Based Precision Approach Local Area Augmentation System (LAAS)—Signal-in-Space Interface Control Document (ICD) [Electronic resource]. Radio Technical Commission for Aeronautics
16. Sokolov AA (2005) Estimation of the LAAS differential correction errors in aviation. Abstract of thesis and technical sciences. S.-Pb (in Russian)
17. Sokolov AI (2006) Monitoring phase measurements in reference receivers of ground local segmentation systems GPS/Glonass. In: Sokolov AI, Chistyakova SS (eds) Publ. House SPbGETU "LETI"/SPbGETU "LETI", 2006. Issue 2: radioelectronics and telecommunications, pp 27–32 (in Russian)
18. Chistyakova SS (2007) Study of the effect of phase measurements jumps on errors in determining the aircraft coordinates in the instrument differential satellite-based landing system. In: Proceedings of the conference 62nd scientific and technical conference on the radio day. Apr 2007. St. Petersburg. Publ. House SPBGETU "LETI" (in Russian)
19. Sauta OI, Sauta AO, Chistyakova SS, Yurchenko YuS, Sokolov AI, Sharypov AA (2011) Effect of the multipath signal propagation on measurement errors in the global navigation satellite system. In: Fourth All-Russian conference "fundamental and applied position, navigation and time support" (PNT-2011). Theses of reports. Institute of Applied Astronomy, Russian Academy of Sciences, St. Petersburg, pp 224–226 (in Russian)
20. Sauta OI, Sokolov AI, Yurchenko YuS (2009) Estimation of data generation accuracy in the local area augmentation system in aviation. In: Radioelectronics issues. Series RLT-2009-Issue 2, pp 183–193 (in Russian)
21. Integrated complex (in Russian)

Chapter 3
Methods for Improving Flight Efficiency and Safety for Satellite-Based Landing Systems

This chapter considers methods used to solve functional problems and to construct devices, the application of which in the SLS allows enhancing the accuracy of determining the parameters of the AC state vector, and improving integrity and continuity of navigation information. The proposed methods implement those most relevant areas of development indicated in the conclusions to the previous chapter. A detailed description of the methods and devices under consideration is given in [1–15].

The proposed methods and devices solve both the problems associated with the decrease in the accuracy of position determination caused by the multipath propagation of the GNSS radio signals and the problem of increasing the interference immunity, including the presence of radiointerference.

3.1 Method for Increasing the Accuracy and Integrity of the Guidance Signals Based on the Construction and Use of Volumetric Distribution Diagrams for Radio Waves Multipath Errors and the System Structure for Its Implementation

Below is described a method for determining the variance of the pseudorange (PR) measurement error in the ground-based and onboard SLS equipment, taking into account the effects of multipath propagation of radio signals. A feature of the proposed method is the formation of volumetric error distribution diagrams based on correction and prediction algorithms. The possibility of extrapolating this method for landing systems of other types is also investigated.

A necessary functional element of the SLS is the device for estimating the integrity of the radio navigation system, which determines the reliability (probabilistic accuracy) of the aircraft positioning. In estimating the integrity, it is assumed that the LAAS transmits error variances in the differential corrections $\sigma^2_{\text{pr\_gnd},n}(k)$ for each nth SV at the kth discrete time instant, and the onboard SLS equipment determines the

© Springer Nature Singapore Pte Ltd. 2020

Baburov S.V. et al., *Development of Navigation Technology for Flight Safety*, Springer Aerospace Technology, https://doi.org/10.1007/978-981-13-8375-5_3

dispersions of onboard errors of the PR measurements $\sigma^2_{\mathrm{air},n}(k)$. Then, the total variance of the PR measurement errors $\sigma^2_{\mathrm{tot},n} = \sigma^2_{\mathrm{air},n} + \sigma^2_{\mathrm{pr\_gnd},n} + \sigma_i^2 + \sigma_t^2$ is determined, where σ_i^2 and σ_t^2 are additional ionospheric and tropospheric errors [16].

To generate a warning signal to the aircraft crew on the degradation of the coordinate measurement accuracy, the system calculates variances of errors in measurements of the aircraft coordinates and the values of protection levels with the vertical and horizontal deviations from the intended glide path. If the value of the coordinate errors exceeds the threshold values corresponding to the safe landing, the integrity evaluator generates a warning. Thus, the safety of landing depends on the accuracy in the determination of variance errors for the coordinate measurements.

The main sources of measurement errors in the differential navigation system are fluctuation interference and multipath propagation of signals manifested both on the LAAS and onboard the aircraft.

To monitor the level of fluctuation interference, a signal-to-noise ratio-meter can be used that is included in the user's SLS equipment.

At present, several methods for GNSS-based landings are known [17, 18], the principle of which is that with the use of the principles of differential navigation [17, 19], the aircraft coordinates are refined and its deviation from the intended landing glide path is determined.

A common drawback of these methods is the incorrect consideration of the ground and onboard errors in measurements due to multipath propagation of radio signals (multipath error), which is one of the reasons for the decrease in the accuracy of determining navigation parameters. The multipath effect arises due to the re-reflection of radio signals from various objects located on the ground in the vicinity of the LAAS reference receiver antenna installation sites. Onboard the aircraft, the reflections come from its structural elements: the fuselage, wings, etc.

Figure 3.1 shows the generalized SLS diagram and some paths of radio signals propagation between SVs and GNSS receivers.

The need to take into account the multipath errors (MP) on both the LAAS and onboard the aircraft is noted in a number of regulatory documents [20, 21], but they do not suggest a specific method for determining this error on the LAAS; however, they suggest that the onboard MP error should be taken into account in such a way that the error estimate used is certainly much higher than the real one [16]. Therefore, there is an actual problem of determining and predicting the variance of the errors caused by the multipath propagation of signals on the LAAS and onboard the aircraft.

Essentially, the proposed method involves increasing the integrity and accuracy of the SLS data based on the formation and use of volumetric distribution diagrams for the predicted variance of the PR definition error.

Figure 3.2 schematically represents the conditional normalized sections of the diagrams of the predicted variance of the ground (GME) and onboard (OME) MP errors. Dotted circles with centers at the points of the LAAS and the onboard antenna installation conditionally illustrate the traditionally used notion of the error variance magnitude.

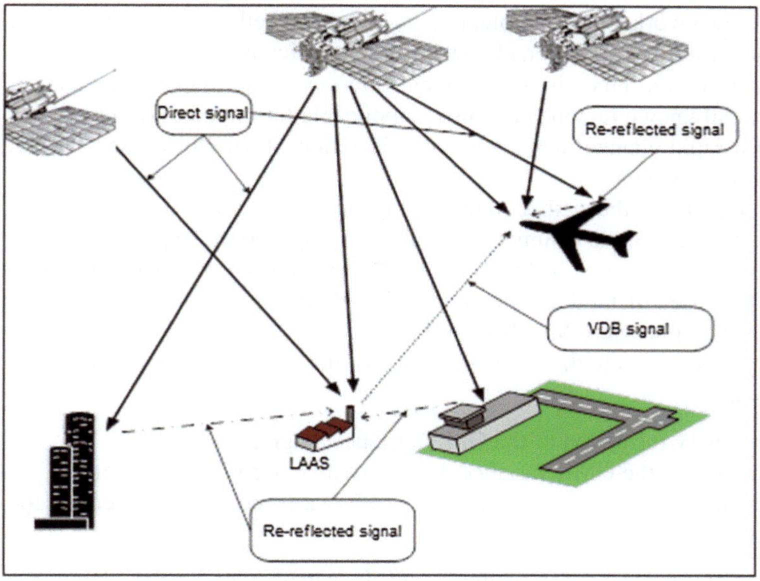

Fig. 3.1 Generalized diagram of the satellite-based landing system and paths of direct and re-reflected radio signals

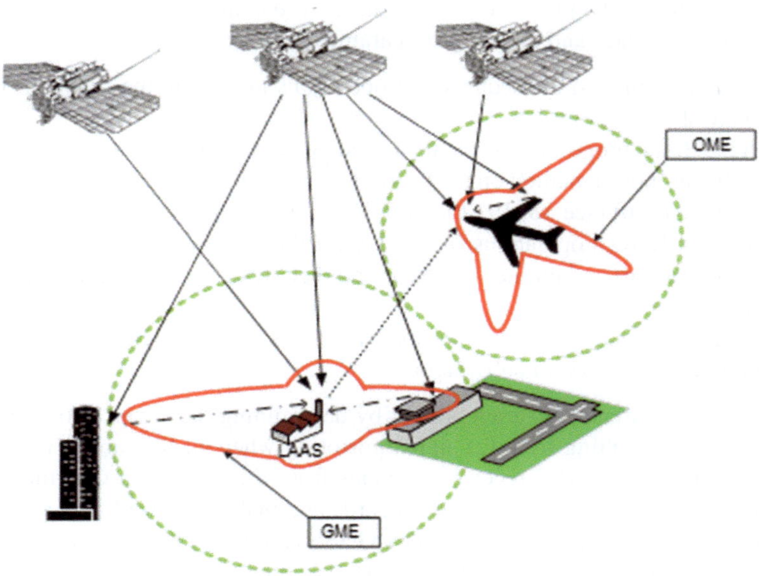

Fig. 3.2 Diagrams of the volumetric distribution of the predicted variance of the ground (GME) and onboard (OME) multipath errors

In known works, for example, [22–24], various methods for determining the MP error are proposed. However, all of them are inapplicable in the SLS because of either short correlation intervals or inefficient filtering of the ionospheric delay of radio signals. All known methods are also inapplicable for onboard MP error evaluations due to a rapid change in the angular orientation of satellites when the aircraft is performing a maneuver.

Let us consider the method of increasing the accuracy and integrity of the SLS data based on the use of the volumetric distribution diagrams of the predicted variances of the ground and onboard MP errors. The basic elements ensuring the implementation of the proposed method are the variance diagrams (scatter diagrams) of the ground (GME) and onboard (OME) MP errors [6].

The main stages in the implementation of the method under consideration at the LAAS are:

- preliminary formation of the GME variance diagram,
- correction of the GME variance diagram with current measurements,
- calculation of the ground PR error (GPE) variance, i.e., differential correction error.

The main stages in the implementation of the method under consideration onboard the aircraft are:

- preliminary formation of the OME variance diagram,
- determination of the PR error variance and its use in the calculation of the refined aircraft coordinates and the integrity parameters.

The relationship and the sequence of operations in solving the problems of this method are shown in Fig. 3.3.

In Fig. 3.3, the proposed new elements and operations that ensure the implementation of the proposed method on the LAAS (see 8, 9, 10, 13, 16, 17 in Fig. 3.3b) and onboard the aircraft (see 18, 19, 20, 24 in Fig. 3.3a) are marked with a rectangular solid contour. Known operations (necessary for the implementation of the method onboard the aircraft and on the LAAS) are marked by an oval or rectangular dotted contour.

Let us consider specific operations that are necessary to implement the method in question in the SLS *ground* subsystem.

(1) The GME variance diagram is built by determining the errors in the PR measurements resulting from the signal mirroring when all SVs are in view at the location of the GNSS receiver antennas from the LAAS. In determining the errors, the differential correction records are analyzed for all SVs using both single-frequency and two-frequency GNSS receivers [20, 21, 24]. In this diagram, the repeatability of the SV paths over time and the correlation of errors and values of the signal-to-noise ratio at the receiver output are used.

To accumulate information, when constructing the GME variance diagram, errors shall be measured over the course of several days. At the same time, the introduction of such information into the LAAS database makes it possible to predict the GME

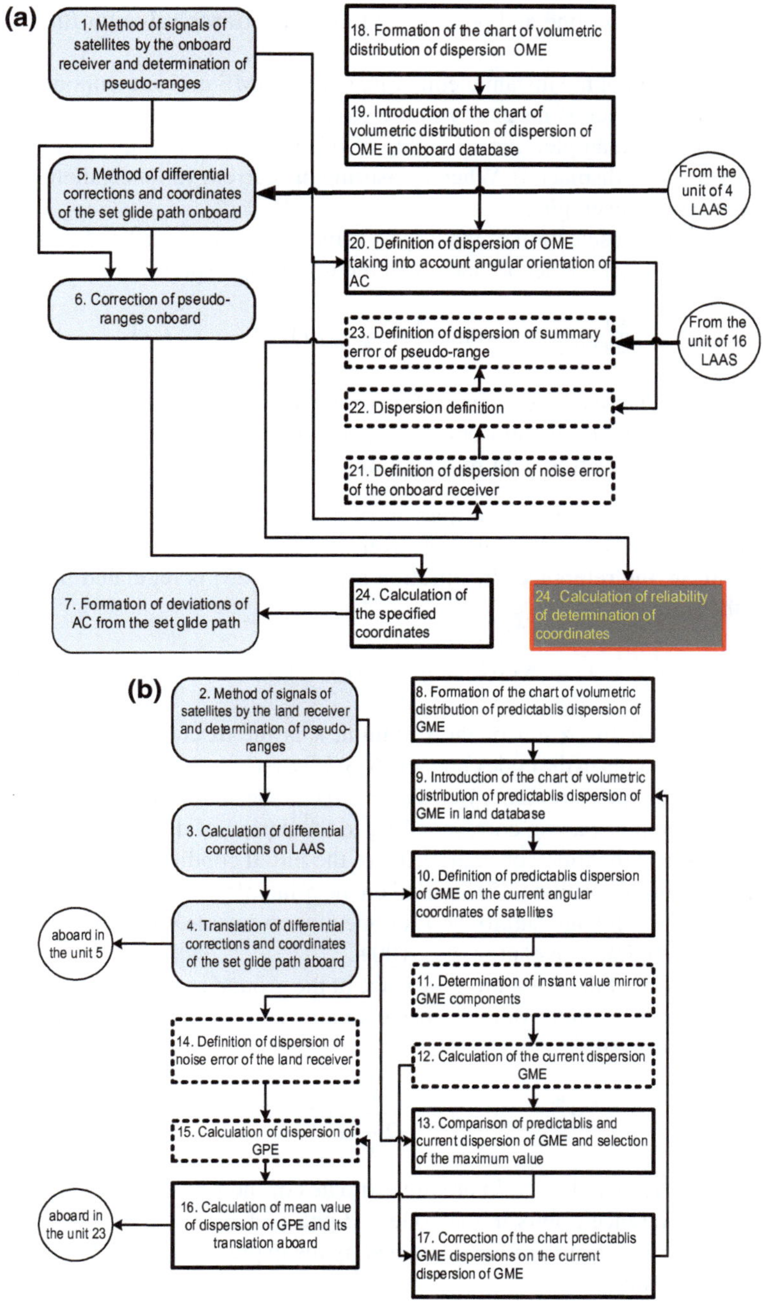

Fig. 3.3 Relationship of functional tasks in the onboard (**a**) and ground (**b**) subsystems of the satellite-based landing system

variance at any given time and produce a predicted variance value for *i*th SV ($i = 1 \ldots N$).

The calculation of the instantaneous value of the GME mirror component and the current GME variance is necessary in the case of rapid changes in the interference environment (for example, during flights of other aircraft or when large objects in the LAAS area are displaced). When measuring the current variance, there is no data accumulation and averaging.

Comparison and selection of the largest values of the predicted and current GME variances $\hat{\sigma}^2_{\text{tropo\_gnd}}(i)$ in accordance with the following expression:

$$\hat{\sigma}^2_{\text{tropo\_gnd}}(i) = \max\left\{\sigma^2_{\text{tropo\_gnd}}(i), \hat{\sigma}^2_{\text{tropo\_bgnd}}(i)\right\} \quad (3.1)$$

makes it possible to exclude the use of incorrect data on differential corrections, for example, when new (movable) objects appear in the LAAS area that significantly affect the overall situation with re-reflections.

(2) The noise error variance on the LAAS is calculated using the signal-to-noise value measured by the ground receiver.
For SLSs, the reference PR filtration algorithm [25] is regulated by equations of the form:

$$y(k) = y(k-1) + \Delta y_2(k) + \alpha \cdot (y_1(k) - y(k-1) - \Delta y_2(k)),\ y(0), \quad (3.2)$$

where $y(k)$ is the result of the filtration; k is the discrete time; $y_1(k)$ are PR reference points in the code channel; $\Delta y_2(k) = y_2(k) - y_2(k-1)$ is the increment of the phase measurements $y_2(k)$; $\alpha = \Delta t/T$ is the weighting factor; $\Delta t = 0.5$ s is time sample spacing corresponding to the period of differential data output; $T = 100$ s is the time constant; $y(0)$ are the initial conditions.
The filtering algorithm (3.2) is based on the principle of invariance to the dynamics of the PR change and to the PR code measurements $y_1(k)$ and is a low-pass filter with a time constant of $T = 100$ s.

(3) The estimate of the noise error variance of the ground receivers $\sigma^2_{\text{noise\_gnd}}(i, k)$ is defined as the estimate of the error fluctuation component variance at the output of this filter [26]:

$$\sigma^2_{\text{noise\_gnd}}(i, k) = \frac{\alpha}{2-\alpha}\sigma_1^2(i, k) + \sigma_2^2(i, k), \quad (3.3)$$

where $\sigma_1^2(i, k)$ and $\sigma_2^2(i, k)$ are the estimates of the variances of noise errors in the PR reference points in the code and phase channels; $\alpha = 5 \times 10^{-3}$ is the weighting factor of the filter described by the expression (3.2); k is the discrete (sampled) time.

The values $\sigma_1^2(i, k)$ and $\sigma_2^2(i, k)$ appearing in (3.3) are determined by the known method [27]:

$$\sigma_1^2(i,k) = \chi_1 \cdot c^2 \cdot \frac{T_0 \cdot T_D \cdot B_C}{2 \cdot Q(i,k)}, \tag{3.4}$$

$$\sigma_2^2(i,k) = \chi_2 \cdot c^2 \cdot \frac{B_P}{(2 \cdot \pi \cdot f_{L1})^2 \cdot Q(i,k)}, \tag{3.5}$$

where χ_1 and χ_2 are rates of loss of energy in the code and phase channels of the satellite signal receiver, respectively; c is the speed of light; T_0 is the length of the selector pulse; T_D is sampling (discretization) period; B_C and B_P are effective bands of the code and carrier phase delay tracking systems, respectively; f_{L1} is L1 carrier frequency; $Q\ (i,\ k)$ is the signal/noise ratio in the 1 Hz bandwidth; k is the discrete (sampled) time.

(4) After determining the GME variance $\hat{\sigma}^2_{\text{tropo\_gnd}}(i,k)$ (see 13 in Fig. 3.3a) and the noise error variance $\sigma^2_{\text{noise\_gnd}}(i)$, the variance of the PR ground error $\sigma^2_{\text{gnd}}(i)$ is calculated:

$$\sigma^2_{\text{gnd}}(i) = \hat{\sigma}^2_{\text{tropo\_gnd}}(i) + \sigma^2_{\text{noise\_gnd}}(i). \tag{3.6}$$

The RMS value of the GRE variance is transmitted to the aircraft over the VDB data link channel that is part of the SLS.

(5) At the same time, the correction of the predicted GME variance diagram (scatter chart) (see 17 in Fig. 3.3) with the current value of the GME mirror component variance is necessary because of the change in the reflecting properties of ground objects when weather and season change. In the process of correction, the known α-filter algorithm is used [23]:

$$\sigma^2_{\text{trropo\_bgnd}}(i,k) = \sigma^2_{\text{tropo\_bgnd}}(i,k-1) + \alpha\Big[\hat{\sigma}^2_{\text{tropo\_gnd}}(k) - \sigma^2_{\text{tropo\_bgnd}}(i,k-1)\Big], \tag{3.7}$$

where α is the weighting factor characterizing the desired ground-diagram updating rate; i is the SV number; k is the discrete (sampled) time for updating the database.

The repeatability of the multipath error in the LAAS is shown in Fig. 3.4 for one of the GPS satellites. In this experiment, the error was formed as the difference between the code and phase measurements. The constant component of the error was eliminated by a high-pass filter. Figure 3.5 shows the change in the elevation angle of this satellite.

Similar measurements for GLONASS satellites are shown in Figs. 3.6 and 3.7. The greater scatter in the values of the errors presented here is due to the fact that different satellites belonging to the same orbit were compared.

The method used to form diagrams can be extended to other instrument landing systems (ground-based beacons, landing radars, etc.). The application of the experi-

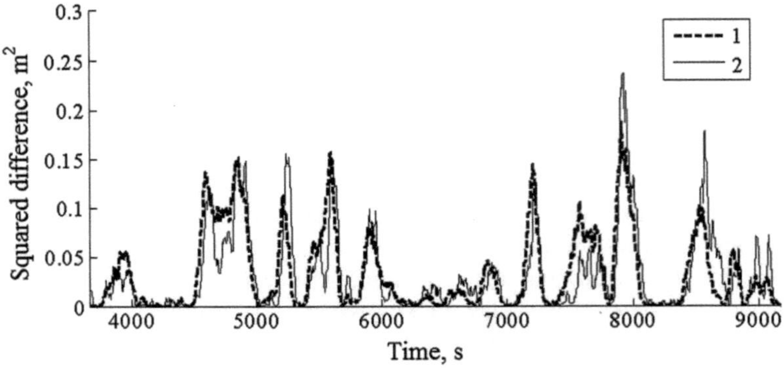

Fig. 3.4 Repeatability of the squared difference in the code and phase measurements of the pseudorange for GPS satellite No. 31. (1) squared averaged difference of observations on 01.04.09, 08.04.09 and 15.04.09; (2) squared difference on 08.04.09

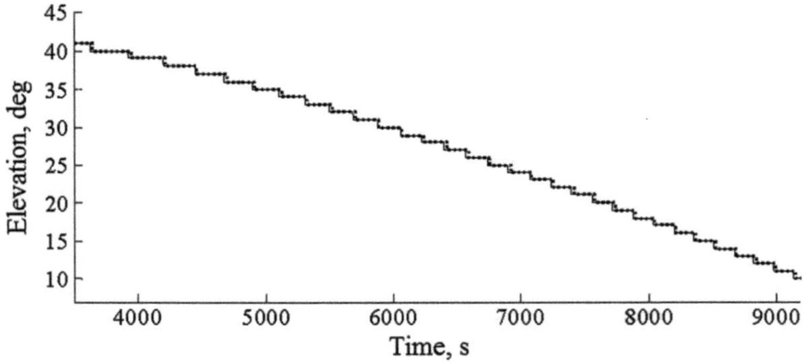

Fig. 3.5 Elevations for GPS satellite No. 31 in observations on 01.04.09 and 08.04.09

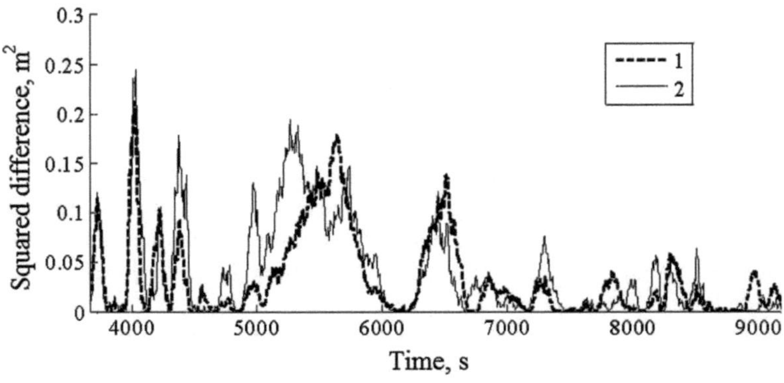

Fig. 3.6 Repeatability of the squared difference in the code and phase measurements of the pseudorange for GLONASS satellite No. 31. (1) squared averaged difference of observations for satellites No. 22 on 01.04.09, No. 21 on 08.04.09 and No. 20 on 15.04.09, (2) squared difference for GLONASS satellite No. 21 on 08.04.09

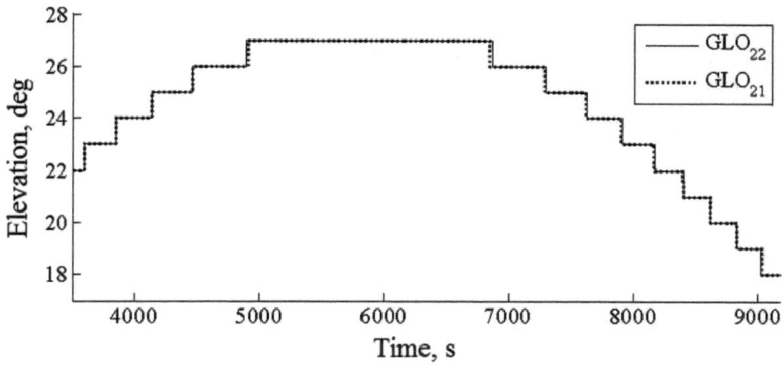

Fig. 3.7 Elevations for GLONASS satellites No. 22 and No. 21

mental method of taking into account the MP effects [28] makes it possible to increase the accuracy of the formation of the diagrams considered.

Efforts have been made to take into account MP effects using the theoretical method for other landing systems, for example, when landing with radio beacons [26]. The combination and analysis of theoretical and experimental methods will allow for the adjustment of volumetric diagrams and further increase the accuracy of their formation.

Now consider the operations (see Fig. 3.3a) that are necessary for the implementation of the method in question in the onboard SLS subsystem.

(1) The volumetric diagram of the MP error variance distribution (OME) (see 18 in Fig. 3.3a) is formed through theoretical or experimental investigation of the scattering diagram regarding the body of the selected aircraft type. Methods for the experimental investigation of signals reflected from aircraft structural elements are known and described, for example, in [28].

In the method under consideration, the results of these reflection studies are first used in the navigation process to form a volumetric diagram of the predicted OME variance value.

The onboard measurement of MP errors in real time, i.e., the determination of the current OME value is hampered due to possible rapid changes in the angular orientation of the SVs relative to the aircraft body during maneuvering.

The current OME variance value (see 20 in Fig. 3.3a) is determined on the basis of the predicted OME variance values, taking into account the angular orientation of the SVs relative to the aircraft body.

To do this, it is necessary to calculate the azimuth and elevation angle of the SVs in the navigation coordinate system (these data are contained in the onboard GNSS receiver) and then, using the measurements of the angular stabilization system of the aircraft or the inertial navigation system, to determine the angular position of the satellites in the coordinate system associated with the aircraft, and the variance value $\sigma^2_{\text{tropo\_air}}(i, k)$ for the ith SV at the kth moment of time.

(2) The variance of the noise error $\sigma^2_{\text{noise\_air}}(i, k)$ of the onboard receiver is determined similarly to the noise error of the ground receiver by the Formula (3.3).
(3) The OPE variance of the ith satellite (see 22 in Fig. 3.3a) is determined by summation of the variances

$$\sigma^2_{\text{air}}(i, k) = \sigma^2_{\text{tropo\_air}}(i, k) + \sigma^2_{\text{noise\_air}}(i, k). \tag{3.8}$$

(4) Then, the variance of the total system error is determined (see 23 in Fig. 3.3) $\sigma^2_{\text{tot}}(i)$ by summation of the variances of the LAAS measurement $\sigma^2_{\text{pr\_gnd}}(i)$, onboard $\sigma^2_{\text{air}}(i, k)$ and additional tropospheric σ^2_{trop} and ionospheric σ^2_{iono} errors recommended in [21],

$$\sigma^2_{\text{tot}}(i) = \sigma^2_{\text{pr\_gnd}}(i) + \sigma^2_{\text{air}}(i, k) + \sigma^2_{\text{iono}} + \sigma^2_{\text{tropo}}. \tag{3.9}$$

Data on measurement errors are used to calculate the refined coordinates and the reliability of their determination (see 24 in Fig. 3.3a).

Operations necessary to implement the proposed method are described in detail in [5, 6].

The loss of *integrity* of the navigation measurements is monitored during landing by comparing the protection levels with the vertical and lateral alert limits. The protective levels are calculated with the $\sigma^2_{\text{tot}}(i)$ value. An inaccurate determination of $\sigma^2_{\text{tot}}(i)$ may cause a false integrity loss alarm. Thus, the errors in determining the value of errors in measuring the coordinates lead to incorrect formation of alarms.

The result of this monitoring is used to warn the crew if the errors in measuring the coordinates exceed the dimensions of the "safety tunnel" where the landing aircraft should be [16].

In order to generate an onboard warning signal according to [20], the variances of the coordinate measurement errors are calculated, as well as the protection levels for the height V and the lateral deviation L from the established glide path. A warning is generated if the protection levels exceed thresholds V_{p} and L_{p} determined by the safety standards.

In the onboard computer, a rectangular right-handed coordinate system is used to provide a navigation solution and to solve integrity monitoring tasks, the origin of this coordinate system is aligned with the approach end of the runway; the Ox-axis is directed along the runway, tangent to the surface of the reference ellipsoid; the Oz-axis is perpendicular to this surface and is directed outside the ellipsoid; and the Oy-axis complements the coordinate system to the right-handed one.

The linearized observation model used for the navigation solution has the following form [20]:

$$\Delta\rho = H \cdot \Delta X + e, \tag{3.10}$$

where $\Delta\rho$ is the vector of deviations of the measured pseudoranges relative to the calculated pseudoranges determined for the given reference point X_0; H is the matrix

Table 3.1 Differential correction error

SV No.	03	05	17	24	28	47	49	51
σ_{tot} (m)	0.3	0.1	0.2	0.4	0.4	0.6	0.8	0.7
σ^*_{tot} (m)	1.2	0.6	1.2	1.1	1.4	0.9	1.3	1.5
F_{Acc}	4.0	6.0	6.0	2.7	3.5	1.5	1.6	2.1

of the direction cosines in the given coordinate system; ΔX is the required coordinate increment vector relative to the reference point X_0; e is a residual (nullity) vector.

The protection levels L and V are calculated as follows:

$$L = k_0 \cdot \sqrt{\sum_{n=1}^{N} S_{2,n}^2 \cdot \sigma_{tot,n}^2}, \tag{3.11}$$

$$V = k_0 \cdot \sqrt{\sum_{n=1}^{N} \left(S_{3,n} + S_{1,n} \cdot \mathrm{tg}(\Theta)\right)^2 \cdot \sigma_{tot,n}^2}, \tag{3.12}$$

where N is the number of SVs used in the navigation solution, $S = [H^T W^{-1} H]^{-1} H^T W^{-1}$ is a projection matrix; $W = \mathrm{diag}\left(\sigma_{tot}^2(1), \ldots, \sigma_{tot}^2(N)\right)$ is a diagonal weight matrix; $S_{i,n}$ is an element of the matrix **S** located in the ith row and nth column; Θ is the glide path angle; k_0 is the coefficient providing the required value of the integrity loss risk.

When calculating the probability of integrity loss, the normal distribution of errors is used. The probability of exceeding the threshold and triggering the warning is given by the formula:

$$Q(x_i) = \frac{2}{\sqrt{2\pi}} \int_0^{x_i} \exp(-t^2/2)\mathrm{d}t, \tag{3.13}$$

where x_1 and x_2 are the ratios of the threshold to the root-mean-square error value in the vertical and lateral channels (planes).

The probability of loss of integrity for the SLS equal to 1×10^{-7} corresponds to a value of 5.327 σ_{tot} when the error value is precisely known.

The increase in the accuracy of determining differential corrections using the method described above and a pre-created GME database is illustrated in Table 3.1. The presented results are obtained by semi-realistic simulation and processing of measured pseudorange data from the onboard GNSS receiver taking into account (σ_{tot}) and without (σ^*_{tot}) of the pre-formed GME characteristics. Table 3.1 also shows the value of the $\mathbf{F}_{\mathbf{Acc}}$ factor (the relative increase in the differential correction accuracy) for the simulation.

Taking into account the fact that the errors in the determination of differential corrections are transformed into coordinate (position) errors with a coefficient proportional to the geometric factor [15], it can also be said that the errors in the determination of the aircraft coordinates will also increase in proportion to the $\mathbf{F}_{\mathbf{Acc}}$. As additional studies show, the mean value of the $\mathbf{F}_{\mathbf{Acc}}$ factor is 3.5.

The results of the semi-realistic simulation using real records obtained during flight tests of the landing system showed that the use of the proposed method makes it possible to improve the accuracy of determining errors in differential corrections by 1.5–6.0 times.

If we assume that for the error σ_{tot} calculated taking into account the GME compensation according to the method considered, the probability of false alarm is 1×10^{-7} [16], then using the error value σ_{tot}^{*}, which is 1.5–6.0 times higher than this value (see Table 3.1), the probability of false alarm will increase to 4×10^{-4} (the upper limit in (3.13) is $5.327/1.5 = 3.55$, which in accordance with (3.13) gives the result 0.0996) or up to 0.38 (the upper limit in (3.13) is $5.327/6 = 0.88$, which in accordance with (3.13) gives the result 0.62), respectively. It is plain, in the latter case, that the normal operation of the SLS is violated due to frequent alarms.

The developed method helps refine the values of general error variances $\sigma_{\text{tot}}^{2}(i)$ by Formula (3.9), which is used to generate the weight and projection matrices in determining the refined coordinates and the integrity of their determination (see 24 in Fig. 3.3a) in accordance with the protection levels calculated by Formulas (3.11) and (3.12).

To increase the reliability of the SLS ground subsystem operation (LAAS), it is advisable to use several diversity receiving channels [21]. In this case, the LAAS differential corrections are formed as the arithmetic mean of the corrections generated by each receiver.

Due to the significant distance between the receivers, the MP errors in these channels are not correlated. Therefore, each receiver shall have its own volumetric diagram of the predicted GME variance.

When using M ground receivers, operations corresponding to the scenario of one ground receiver are performed for each of them (see Fig. 3.3b).

After selecting the maximum value, the MP error variance when receiving the ith satellite's signal in the kth receiver is equal to $\widehat{\sigma}_{\text{trropo\_gnd}}^{2}(i,k)$, and after calculating 14 (see Fig. 3.3), the variance of the noise component of the PR measurement error is obtained. Next, the GPE variance is calculated for the kth receiver:

$$\sigma_{\text{gnd}}^{2}(i,k) = \sigma_{\text{tropo\_gnd}}^{2}(i,k) + \sigma_{\text{noise\_gnd}}^{2}(i), k. \tag{3.14}$$

Then, the GPE RMS value of the ith satellite is determined as the square root of the sum of the GPE variances of each ground receiver divided by the number of receivers:

$$\sigma_{\text{gnd}}(i) = \frac{1}{M} \cdot \sqrt{\sum_{k=1}^{M} \sigma_{\text{gnd}}^{2}(i,k)}. \tag{3.15}$$

The resultant GPE value is transmitted to the aircraft.

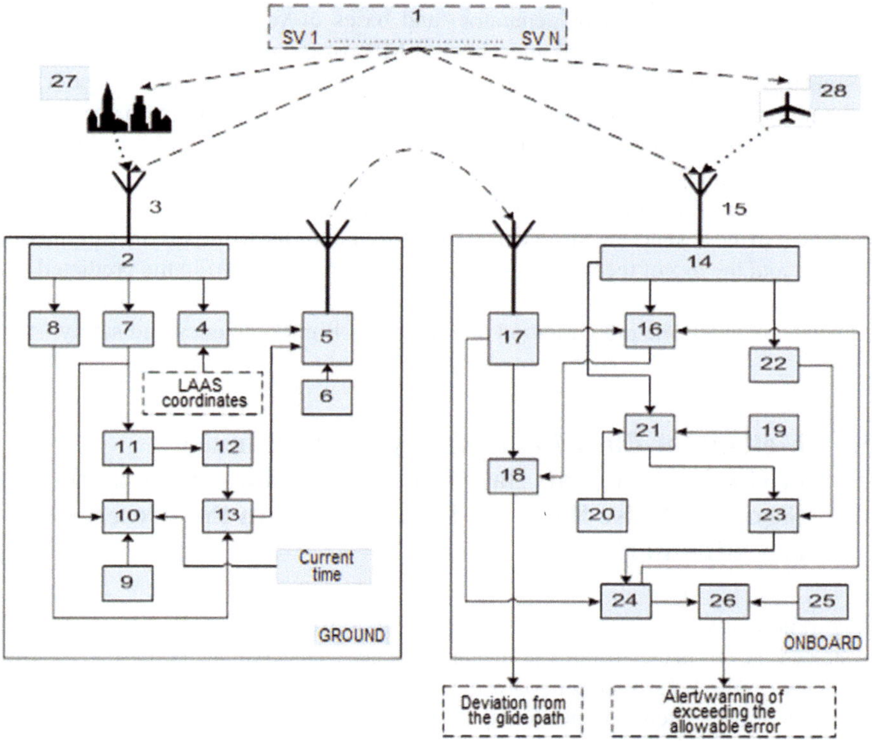

Fig. 3.8 SLS structure for implementing the method of using the volumetric distribution diagrams for radio waves multipath errors. The numbers indicate: 1—Constellation of navigation satellites. 2—Ground receiver. 3—Ground antenna for satellite signals. 4—Differential correction calculator. 5—Data link transmitter with antenna. 6—Ground database of the intended glide path. 7—Calculator of the current ground MP error variance. 8—Calculator of the ground receiver noise error variance. 9—Generator of the volumetric diagram of the predicted ground MP error variance. 10—Database of the predicted ground MP error variance. 11—Comparator. 12—Determinant of the maximum ground MP error. 13—Calculator of the root-mean-square value of the ground PR error. 14—Onboard receiver. 15—Antenna of the onboard receiver for satellite signals. 16—Calculator of the current aircraft coordinates. 17—Data link receiver with antenna. 18—Calculator of deviations from the intended glide path. 19—Database of the predicted onboard MP error variance. 20—Determinant of the aircraft attitude. 21—Calculator of the onboard MP error variance. 22—Calculator of the onboard receiver noise error variance. 23—Calculator of the onboard PR error variance. 24—Adder (summator) of PR errors. 25—Onboard database of allowable errors in determining the aircraft coordinates. 26—Calculator of the aircraft coordinates reliability. 27—Reflecting objects in the LAAS area. 28—Re-reflecting elements of the aircraft structure

The SLS structure that implements the proposed method is shown in Fig. 3.8. In addition, Fig. 3.8 shows the reflecting objects in the area of the LAAS location (27) and re-reflecting elements of the aircraft structure (28). The operation of devices implementing the proposed SLS structure is described in detail in [5, 6].

Thus, the introduction of generators and bases of volumetric diagrams of the predicted variances of ground and onboard MP errors and their use in the landing system together with known units that determine and process differential pseudorange corrections and the introduced blocks calculating variances of multipath propagation and noise errors make it possible to refine the aircraft coordinates (position) and increase their integrity.

In addition, as mentioned above, the continuity of the SLS ground subsystem is increased in the absence of current GME measurements due to the variance prediction channel and the use of the maximum GME value when comparing the predicted and current variances.

Next, consider one more method that increases flight efficiency and safety.

3.2 Method for Ensuring Integrity and Continuity of Guidance Signals Based on the Use of an Integrated Signal-to-Noise Ratio for Pseudoranges in the Presence of Radiointerference

A common drawback of the known methods of constructing satellite-based landing systems [29, 30] is the absence of means to detect unacceptable decrease in the SLS accuracy on the LAAS and onboard the aircraft. The reason for the decrease in the accuracy of the system operation may be emissions of signals from third-party radio technical systems near the LAAS receiver antennas or onboard antennas (the effect of radiointerference of these signals and SV signals). Given the high sensitivity of GNSS receivers, not only the fundamental frequency of a third-party system radiation, but also its harmonics can be dangerous.

Figure 3.9 shows the propagation paths of radio signals from the ground source of interference and the onboard transmitter of a satellite communication system of the INMARSAT type.

The need to monitor the accuracy of the LAAS operation is indicated in a number of documents (for example, [21]), but they do not offer any particular method for detecting unacceptable decrease in the LAAS operation accuracy due to the radiointerference effects. These documents also indicate the need to monitor the SLS integrity. To monitor the integrity onboard the aircraft, it is necessary to have information about the "Ground Continuity and Integrity Designator," [21, 31] which characterizes the quality of the LAAS operation in the presence of radiointerference effects.

All the known works in this field [19, 28, 32, etc.] lead either to a significant complication and increase in the cost of the LAAS and onboard GNSS/LAAS equipment, or they do not allow the integrity monitoring to be performed in a timely manner.

The main disadvantage of the known methods is the lack of operations related to the detection of unacceptable decrease in the SLS accuracy due to the radiointerference effects, and the insufficiently high accuracy of estimating the ground MP error

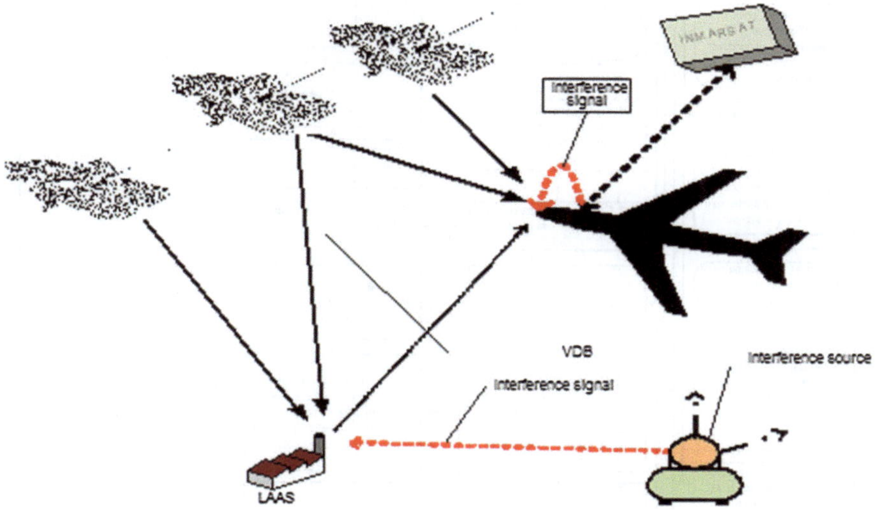

Fig. 3.9 Paths of interference signals causing radiointerference with GNSS signals

due to the failure to take account of the MP effects with large delays of re-reflected radio signals and ionospheric effects. Moreover, when estimating the noise error variance (on the LAAS and onboard the aircraft), the diffusion component of the MP error and the MP effects on the measurement of the signal-to-noise ratio are not taken into account. The elimination of these drawbacks is addressed in papers [5–8].

The task of the proposed new method is to increase the integrity of the aircraft positioning due to the detection unacceptable decrease in the accuracy of the LAAS operation due to the radiointerference effects. At the same time, the current values of the ground MP error (GME) variance and the noise error variance (on the LAAS and onboard the aircraft) are estimated more accurately in comparison with the method described in Sect. 3.1.

The principle of the claimed method is explained with the help of Fig. 3.10 and is described in detail in [7, 15]. The proposed new operations in Fig. 3.10 are highlighted in gray.

The new operations are as follows.

The LAAS generates a time distribution diagram for the predicted GISR; the diagram is entered into the ground database; and the predicted GISR value is determined (see block 13 in Fig. 3.10). For the generation of the time distribution diagram of the predicted GISR, preliminary measurements of ground signal-to-noise ratios made under normal interference conditions are used to calculate the weighted average of the signal-to-noise ratio for all N visible SVs at the given time, using the following Formula [33]:

$$\overline{\mathrm{SNR}}_{\mathrm{pred}}(t_k) = \frac{\sum_{i=1}^{N} \alpha_i(t_k)\mathrm{SNR}_i(t_k)}{\sum_{i=1}^{N} \alpha_i(t_k)}, \tag{3.16}$$

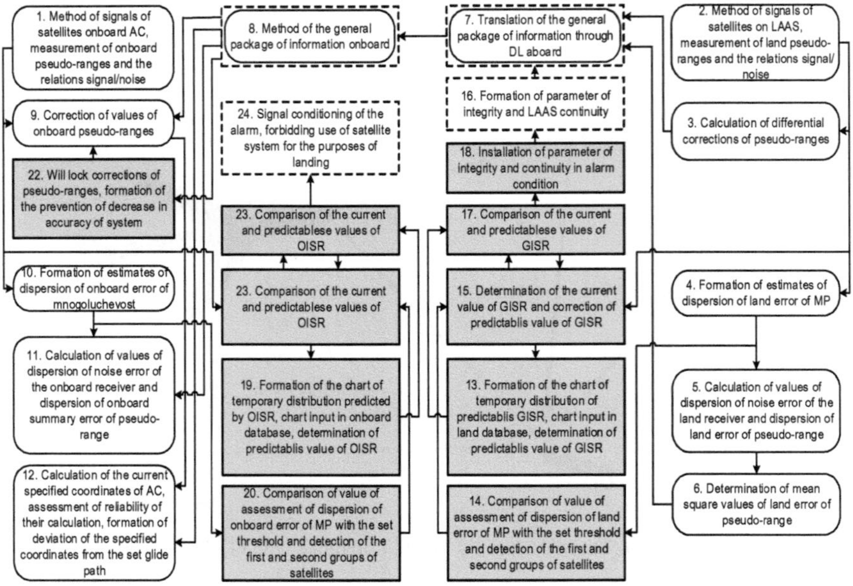

Acronyms: GISR is a ground (on the LAAS) integrated signal-to-noise ratio in the reference receiver, OISR is a predicted onboard (in the GNSS/LAAS equipment) integrated signal-to-noise ratio

Fig. 3.10 Method of ensuring the information integrity and continuity in the satellite-based landing system

where $\mathrm{SNR}_i(t_k)$ is the ground signal-to-noise ratio for the ith SV at a time t_k; $\alpha_i(t_k)$ is weighting factor; $\overline{\mathrm{SNR}}_{\mathrm{pred}}(t_k)$ is the ground integrated signal-to-noise ratio (GISR), further considered as the predicted GISR value.

The use of weighting factors $\alpha_i(t_k)$ allows elimination of the jumps depending on the $\overline{\mathrm{SNR}}_{\mathrm{pred}}(t_k)$ value after the SV constellation changes, when the number of terms N in the weight sum changes.

To do this, the functional dependence of the weighting factor of the ith SV on the elevation angle $\mathrm{EL}(t_k)$ of this SV is applied, which ensures "soft" inclusion (or elimination) of this SV in the GISR calculation operation.

As an example of such a functional dependency, the following dependence can be given:

$$\alpha_i(t_k) = \begin{cases} 1, \text{ if } \mathrm{EL}(t_k) \geq 12, \\ \frac{1}{12 - \mathrm{EL}(t_k)} \end{cases}, \text{ if } \mathrm{EL}(t_k) < 12. \tag{3.17}$$

The results of calculating $\overline{\mathrm{SNR}}_{\mathrm{pred}}(t_k)$ are further averaged using the effects of the repeatability of the angular position of the navigation system SV. Then, GISR values are entered into the database and, in the future, the predicted GISR value is determined from this database (see block 13 in Fig. 3.10) in accordance with the current time and the SV constellation used (if any SV is removed from the number of operating SVs and replaced by a backup one, then the GISR value is calculated with the parameters of the new (formerly backup) SV).

The value of the GME variance estimate is compared with the predetermined threshold and the first and second SV groups are identified (see block 14 in Fig. 3.10). For the first SV group, the value of the GME variance estimate is below the threshold and, consequently, the signals of the first SV group are not subject to the MP effects. The threshold is selected based on the value of the root-mean-square error (noise component) σ_{GME} in estimating the GME in the least favorable (with a small $\text{EL}(t_k)$ value) operating conditions of the channel SV (for example, the threshold can be set to $3\sigma_{\text{GME}}$).

Using the measurements of the ground signal-to-noise ratio for the first SV group the signals of which are not subject to MP effects, the current GISR value is determined (see block 15 in Fig. 3.10). For this purpose, current measurements of the ground signal-to-noise ratio are used to calculate the weighted average of the signal-to-noise ratio $\overline{\text{SNR}}_{\text{curr}}(t_k)$ using the formula given earlier (the only difference is discarding satellite channels in which the GME variance estimate exceeds the predetermined threshold; this discarding is necessary due to distortions in measurements of the signal-to-noise ratio not related to the radiointerference effects). Measurements of the ground signal-to-noise ratio for the second SV group, the signals of which are subject to MP effects, are not used to determine the current GISR value, and the weighted average value of the signal-to-noise ratio $\overline{\text{SNR}}_{\text{curr}}(t_k)$ is considered to be the current GISR value.

At the same time, the LAAS integrity and continuity indicator is formed (see block 16 in Fig. 3.10) (Ground Continuity and Integrity Designator [21, 31]) and transmitted through the DL to the aircraft. In the normal operating mode of the LAAS, this designator is assigned a value of one [21].

Next, the current and predicted GISR values are compared (see block 17 in Fig. 3.10) and, in case of exceeding the predicted GISR values by a given threshold value, the LAAS integrity and continuity designator value is set to the alarm state (see block 18 in Fig. 3.10) (in this case, this value is seven [34]). In the presence of radiointerference effects, the ground signal-to-noise ratio is reduced, which makes it possible to promptly monitor the accuracy of the LAAS operation. The threshold value is selected taking into account the root-mean-square value of the noise component $\sigma_{\text{signal-to-noise}}$ in measurements of the ground signal-to-noise ratio in the least favorable (with a low value of $\text{EL}(t_k)$) under operating conditions of the satellite channel. When the current GISR value is formed, the noise component decreases to a value of $\bar{\sigma}_{\text{signal-to-noise}} = \sigma_{\text{signal-to-noise}}/\sqrt{N}$ as a result of averaging. The threshold is selected in such a way as to achieve a compromise in the integrity and continuity of the navigational measurements (for example, the threshold value can be set to $5\bar{\sigma}_{\text{signal-to-noise}}$).

If the predicted GISR value is not exceeded by the specified threshold value, the predicted GISR value stored in the ground database is adjusted with the current GISR value (see block 15 in Fig. 3.6). This adjustment is required due to the changes in the power of the SV-radiated radio signal and the conditions of its propagation when weather and seasons change. The adjustment involves the known α-filter algorithm [35]:

$$\overline{\mathrm{SNR}}^{+}_{\mathrm{pred}}(t_k) = \overline{\mathrm{SNR}}^{-}_{\mathrm{pred}}(t_k) + \alpha\left[\overline{\mathrm{SNR}}_{\mathrm{curr}}(t_k) - \overline{\mathrm{SNR}}^{-}_{\mathrm{pred}}(t_k)\right], \tag{3.18}$$

where $\overline{\mathrm{SNR}}^{+}_{\mathrm{pred}}(t_k)$ is the previously predicted GISR value stored in the database; $\overline{\mathrm{SNR}}^{-}_{\mathrm{pred}}(t_k)$ is the adjusted predicted GISR value; α is the weighting factor characterizing the desired update rate of the ground diagram of the GISR dependence on time.

Onboard the aircraft, in the same way as on the LAAS, the time distribution diagram is formed for the predicted GISR value; this diagram is entered into the onboard database; the predicted GISR value is determined (see block 19 in Fig. 3.10).

The value of the OME variance estimate is compared with the predetermined threshold and the first and second SV groups are identified (see block 20 in Fig. 3.10). For the first SV group, the value of the OME variance estimate is below the threshold and, therefore, the signals of the first SV group are not subject to the MP effects. When selecting a threshold value, the same considerations apply as on the LAAS.

Using measurements of the onboard signal-to-noise ratio for the first SV groups, the signals of which are not subject to the MP effects, the current OISR value is determined (see block 21 in Fig. 3.10). For this purpose, current measurements of the onboard signal-to-noise ratio are used to calculate the weighted average of the signal-to-noise ratio $\overline{\mathrm{SNR}}_{\mathrm{curr}}(t_k)$ from the formula given earlier.

Measurements of the onboard signal-to-noise ratio for the second SV group, the signals of which are subject to the MP effects, are not used to determine the current OISR value, and the weighted average value of the signal-to-noise ratio is considered as the current OISR value.

Onboard the aircraft, the obtained LAAS integrity and continuity designator is analyzed. If the LAAS operates in the nominal mode (the designator state is equal to one), the LAAS differential corrections received over the DL are used to adjust the pseudoranges measured onboard the aircraft (see block 9 in Fig. 3.10).

If radiointerference effects are detected on the LAAS (the designator state is equal to seven), then the adjustment of the pseudoranges with the differential corrections is prohibited (the differential mode is disabled) and a corresponding warning about the decrease of the system accuracy is output (see block 22 in Fig. 3.10).

In addition, onboard the aircraft, the current and predicted OISR values are compared (see block 23 in Fig. 3.6) and, in case of exceeding the predicted OISR by a predetermined threshold value, an alarm signal is output prohibiting the use of a satellite system for aircraft landings (see block 24 in Fig. 3.10). When selecting the threshold, the same considerations apply as on the LAAS.

The proposals on the method for estimating the LAAS GME variance using code and phase measurements of the ground pseudoranges and signal-to-noise ratios are explained in Fig. 3.11.

The proposed method involves the formation of mid-frequency variation in measurements of the ground signal-to-noise ratio (GSR), formation of mid-frequency

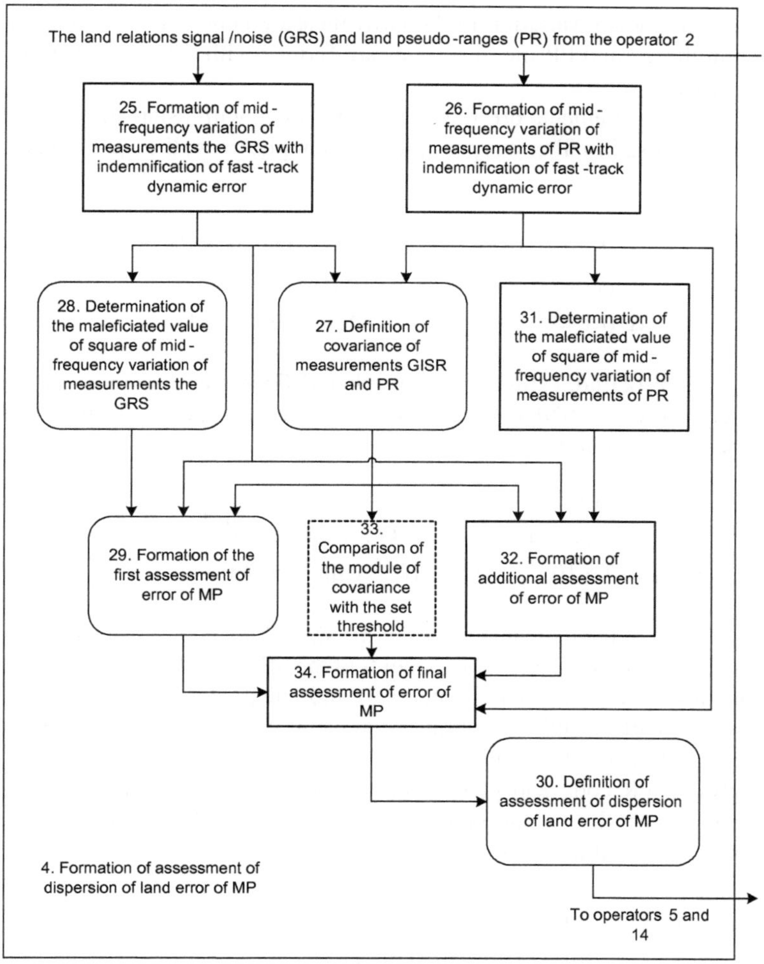

Fig. 3.11 Method of ensuring the information integrity and continuity in the satellite-based landing system

variation of the ground pseudorange (PR) measurements performed on the LAAS, calculation of the product of the mentioned mid-frequency variations and determination of the covariance of the measurements of the GSR and ground PR by smoothing this product, determination of the smoothed square's mid-frequency variation of the GSR measurements, formation of the first MP error estimate by multiplying the mid-frequency variation of the GSR measurements by covariation and dividing the result by the smoothed squared mid-frequency variation of the GSR measurements, estimation of the GME variance by squaring the MP error estimate.

The mid-frequency variation in the GSR measurements and the mid-frequency variation in the ground PR measurements are formed with the compensation of the velocity dynamic error.

The GSR measurements are filtered using a filter with astatism of at least the second order. As such a filter it is possible, for example, to use the $\alpha\beta$-filter algorithm [26]:

$$\begin{aligned} x_1(k) &= x_1(k-1) + \Delta t\, x_2(k-1) + \alpha(k)\Delta_1(k), \\ x_2(k) &= x_2(k-1) + \beta(k)\Delta_1(k), \\ \Delta_1(k) &= z_1(k) - [x_1(k-1) + \Delta t\, x_2(k-1)], \end{aligned} \tag{3.19}$$

where $x_1(k),\ x_2(k)$ are estimates of the filtered GSR measurements and its rate, respectively; $z_1(k) = \mathrm{SNR}(k)$ are GSR measurements; Δt is the time sampling interval; k is the current discrete time;

$$\begin{aligned} &\alpha(k) = \frac{2(2k-1)}{k(k+1)},\ \beta(k) = \frac{6}{k(k+1)\Delta t} \text{ with } k < 3000 \text{ and} \\ &\alpha(k) = 2 * 10^{-3},\ \beta(k) = 10^{-6} \text{ with } k \geq 3000. \end{aligned}$$

The selection of the indicated values of the smoothing coefficients $\alpha(k)$ and $\beta(k)$ provides filtering of the spectral components of the MP error that cannot be estimated by this method, with a correlation interval of more than 500 s.

The selection of variable smoothing coefficients reduces the duration of the transient process of the $\alpha\beta$-filter algorithm.

The ground PR code and phase measurements are filtered using a filter with astatism of at least the second order, and the algorithm of the $\alpha\beta$-filter described earlier should be modified using the invariance principle [26]:

$$\begin{aligned} y_1(k) &= y_1(k-1) + \delta z_3(k) + \Delta t y_2(k-1) + \alpha(k)\Delta_2(k), \\ y_2(k) &= y_2(k-1) + \beta(k)\Delta_2(k), \\ \Delta_2(k) &= z_2(k) - [y_1(k-1) + \delta z_3(k) + \Delta t y_2(k-1)], \end{aligned} \tag{3.20}$$

where $y_1(k),\ y_2(k)$ are estimates of the filtered ground PR and its rate, respectively; $z_2(k)$ are ground PR code measurements; $\delta z_3(k) = z_3(k) - z_3(k-1)$ is increment of the ground PR phase measurements.

As the output signals of $\alpha\beta$-filters, filter discrepancy signals (residuals) $\Delta_1(k)$ and $\Delta_2(k)$ are used, and the $\alpha\beta$-filters themselves act as high-pass filters. Estimates of velocities provide compensation for the velocity dynamic error caused by a change in the GSR average level (in the first $\alpha\beta$-filter), and the ionospheric delay of the radio signal (in the second $\alpha\beta$-filter). As a result, the residuals $\Delta_1(k)$ and $\Delta_2(k)$ do not have any displacements when forming an MP error with a correlation interval of up to 500 s.

The residuals $\Delta_1(k)$ and $\Delta_2(k)$ are then smoothed using algorithms of α-filters with the smoothing coefficients of 0.01 described in [34], so that at the output of these filters, mid-frequency variations in the GSR measurements $\Delta_1^*(k)$ are formed with the compensation of the velocity dynamic error and mid-frequency variations in the ground PR measurements $\Delta_2^*(k)$ with the compensation of the velocity dynamic error with the correlation interval of 100–500 s.

When estimating the MP error using GSR measurements, the following considerations should be taken into account. The connection between the MP error M and the mid-frequency variations $\Delta_1^*(k)$ and $\Delta_2^*(k)$ can be described by expressions [22]:

$$\begin{aligned} \Delta_1^* &= \beta M + \upsilon_1, \\ \Delta_2^* &= M + \upsilon_2, \end{aligned} \tag{3.21}$$

where β is a normalizing factor; υ_1 is a GSR measurement error; υ_2 is ground PR code measurement error (including the noise and ionospheric components).

A specific feature of estimating the MP error $\widehat{M}$ is the impossibility of separating the ionosphere effect and the MP effect in the ground PR measurements of a single-frequency receiver. That is why the estimation $\widehat{M}$ is formed using the GSR measurements, but to do so, it is necessary to find a normalizing factor β. We note, however, that due to the compensation of the velocity error in the $\alpha\beta$-filters, the contribution of the ionosphere effect to the error υ_2 is small, and in the extreme case, statistics $\Delta_2^*(k)$, can also be used as a solution to our problem.

In addition, the smoothed value of the squared mid-frequency variation of the ground PR measurement is determined and an additional estimate of the MP error is generated by multiplying the mid-frequency variation of the GSR measurements by the smoothed squared mid-frequency variation of the ground PR measurements and dividing the result by the covariance mentioned.

The above operations are described mathematically by the following expressions:

$$\begin{aligned} E\{\Delta_1^*\Delta_2^*\} &= \beta E\{M^2\}, \\ E\{(\Delta_1^*)^2\} &= \beta^2 E\{M^2\} + \sigma_1^2, \\ E\{(\Delta_2^*)^2\} &= E\{M^2\} + \sigma_2^2, \end{aligned} \tag{3.22}$$

where $E\{.\}$ is the averaging operator (the averaging or smoothing operation is performed by an α-filter with a smoothing coefficient of less than 0.001); $E\{\Delta_1^*\Delta_2^*\}$ is a covariance of GSR and ground PR measurements; $E\{(\Delta_1^*)^2\}$ is the smoothed value of the squared mid-frequency variation of the NOC measurements; the smoothed value of the square of the mid-frequency variation of the ground PR measurements; σ_1^2 and σ_2^2 are variances of errors υ_1 and υ_2, respectively; a natural assumption is made about the absence of a mutual correlation between all constituents of the components $\Delta_1^*(k)$ and $\Delta_2^*(k)$.

Further, the normalizing factor is estimated:

$$\widehat{\beta}_1 = \frac{E\left\{\left(\Delta_1^*\right)^2\right\}}{E\left\{\Delta_1^* \Delta_2^*\right\}} = \beta + \sigma_3^2 \tag{3.23}$$

and the reciprocal of the normalizing factor:

$$\widehat{\beta}_2 = \frac{E\left\{\left(\Delta_2^*\right)^2\right\}}{E\left\{\Delta_1^* \Delta_2^*\right\}} = \frac{1}{\beta} + \sigma_4^2, \tag{3.24}$$

where σ_3^2 and σ_4^2 are errors in estimating β and $1/\beta$, respectively.

These estimates are used to obtain two MP error estimates:

$$\widehat{M}_1 = \frac{\Delta_1^*}{\widehat{\beta}_1} = \frac{\Delta_1^* E\left\{\Delta_1^* \Delta_2^*\right\}}{E\left\{\left(\Delta_1^*\right)^2\right\}}, \tag{3.25}$$

the first MP error estimate,

$$\widehat{M}_2 = \Delta_1^* \widehat{\beta}_2 = \frac{\Delta_1^* E\left\{\left(\Delta_2^*\right)^2\right\}}{E\left\{\Delta_1^* \Delta_2^*\right\}}, \tag{3.26}$$

an additional MP error estimate. On the other hand, the estimate $\widehat{\beta}_2$ is also inflated, but this leads to inflation of the estimate $\widehat{M}_2$ also.

A compromise solution is proposed: to generate a final MP error estimate, the arithmetic average of the estimates $\widehat{M}_1$ and $\widehat{M}_2$ should be used, i.e.,

$$\widehat{M} = \frac{1}{2}\left(\widehat{M}_1 + \widehat{M}_2\right). \tag{3.27}$$

The estimate of the mid-frequency error component of the mirror MP obtained in this way is based on the assumption of a linear connection between the MP effects in the ground PR code measurements and GSR measurements. Such an assumption is valid in the presence of one strong mirror reflection and a slight delay of the re-reflected signal, leading to distortion of the correlation peak of the PR code. When multiple reflections with different delays occur, the common factor β cannot be found. In the case of MP effects with a long delay of the re-reflected signal, the rear declivity of the correlation peak may be distorted. This causes an error in the PR code measurements but may not affect the GSR measurements that are made at the correlation peak. Such effects are not taken into account in the obtained estimate.

To eliminate this drawback, it is possible to use the obtained MP error estimate only when there is a significant correlation between the ground PR code measurements and the GSR measurements. As the statistics characterizing such a correlation, the covariance of the GSR and ground PR measurements $E\{\Delta_1^* \Delta_2^*\}$ is applied. In this case, the covariance module is compared with a predetermined threshold: $\mathrm{mod}\left(E\{\Delta_1^* \Delta_2^*\}\right) > \mathrm{Th}$. If the threshold is exceeded, a final estimate of the MP error is formed as the arithmetic average of the two mentioned MP estimates, and if the threshold is not exceeded, then, considering the uselessness of the GSR measurements, the mid-frequency variation of the ground PR measurement is used as the final MP error estimate.

When estimating the GME variance on the basis of the instantaneous value of the GME mirror component obtained by the proposed method, the estimate probability density function $\widehat{M}$ is taken into account. According to [15], the distribution of the instantaneous values of a harmonic signal with a random phase follows the arcsine law with variance equal to half the square of the harmonic signal amplitude. Since the protection levels onboard the aircraft are calculated under the assumption of the normal law of PR errors, the real law of the MP error mirror component distribution should be approximated by the normal law. As a result, the estimation of the ground MP error (GME) variance is defined as half the square of the estimate $\widehat{M}$.

The proposals regarding the method for determining the noise error (on the LAAS and onboard the aircraft) are explained below.

In Zaezdny [34], a method is proposed for estimating the receiver noise level based on the dependence of the noise error on the signal-to-noise ratio measured by the navigation receiver. The idea of this method lies in the fact that the delay tracking system error variance (that is, the variation of the PR measurements) at a known signal-to-noise ratio is not difficult to calculate by the formula:

$$\sigma_{\mathrm{PR}}^2 = \lambda_C^2 \frac{d B_{\mathrm{L}}}{2C/N_0}, \tag{3.28}$$

where d is the correlator parameter; B_{L} is one-sided noise band of the delay tracking system; C/N_0 is the ratio of the signal power to the noise power spectral density (measured by the receiver), λ_C is the length of the radio signal element *C/A*-code.

Further, the variance σ_{PR}^2 is recalculated to the output of the code-phase filtration algorithm:

$$\sigma_{\mathrm{Noise}}^2 = \frac{\alpha}{2-\alpha}\sigma_{\mathrm{PR}}^2 \approx \frac{\alpha}{2}\sigma_{\mathrm{PR}}^2, \tag{3.29}$$

where α is the smoothing coefficient of the PR code-phase filtering algorithm; $\sigma_{\mathrm{Noise}}^2$ is the noise error variance.

This method has a number of drawbacks that make it difficult to use in the aircraft landing system, namely:

(1) The MP mirror effects affect the estimation of the signal-to-noise ratio of the ground receiver, which significantly distorts the results of estimating the receiver noise level.
(2) The method does not take into account the diffusion MP effect on the errors in PR measurements.
(3) The method does not take into account the effect of the residual error of the mirror MP with a correlation interval of less than 100 s (this error is suppressed by the code-phase filtering algorithm, but 100% suppression is impossible).

In order to eliminate the above drawbacks, the following is proposed to determine the noise error variance for the ground and onboard receivers.

To determine the variation of the PR measurements, sample variance is generated instead of the calculation, taking into account both the noise component and the diffusion component of the MP error. In this case, as a statistic characterizing the level of error, a residual of the PR code-phase filtration algorithm is used, the square of which Δ_{Φ}^2 is smoothed using a recursive algorithm to obtain sample variance [34]:

$$\sigma_{\text{PR}}^2(i+1) = \sigma_{\text{PR}}^2(i) + \gamma\left\{\Delta_{\Phi}^2(i+1) - \sigma_{\text{PR}}^2(i)\right\}, \quad \sigma_{\text{PR}}^2(0) = 0, \tag{3.30}$$

where γ is the parameter that affects the width of the "sliding window", where the noise error variance is calculated.

When selecting the parameter γ, a compromise solution is used: the smaller the γ value (wider the "sliding window"), the more accurate the estimate σ_{PR}^2, provided the $\Delta_{\Phi}(i)$ process is steady-state; however, when the SV elevation angle θ changes, the steady-state conditions of the process $\Delta_{\Phi}(i)$ are violated and the width of the "sliding window" cannot be increased excessively. It is recommended to select the γ parameter value equal to the smoothing coefficient α of the PR code-phase filtering algorithm.

$$\gamma = \alpha = \frac{\Delta t}{T_{\Phi}} = \frac{0.5}{100} = 0.005,$$

where T_{Φ} is the time constant of the PR code-phase filtration algorithm; Δt is a time sampling interval equal to the period of the differential data repetition in the LAAS message.

The σ_{PR}^2 variation value to the output of the code-phase filtering algorithm (i.e., to the σ_{Noise}^2 value) is converted in accordance with (3.28).

In the proposed method, the variance of the receiver noise error in real time and in combination with the definition of the MP error diffusion component are determined.

In addition, it takes into account the residual error of the mirror MP with a correlation interval of less than 100 s that present in the residual $\Delta_{\Phi}(i)$, eliminates the effect of the mirror MP mid-frequency error on the estimate of the receiver noise error.

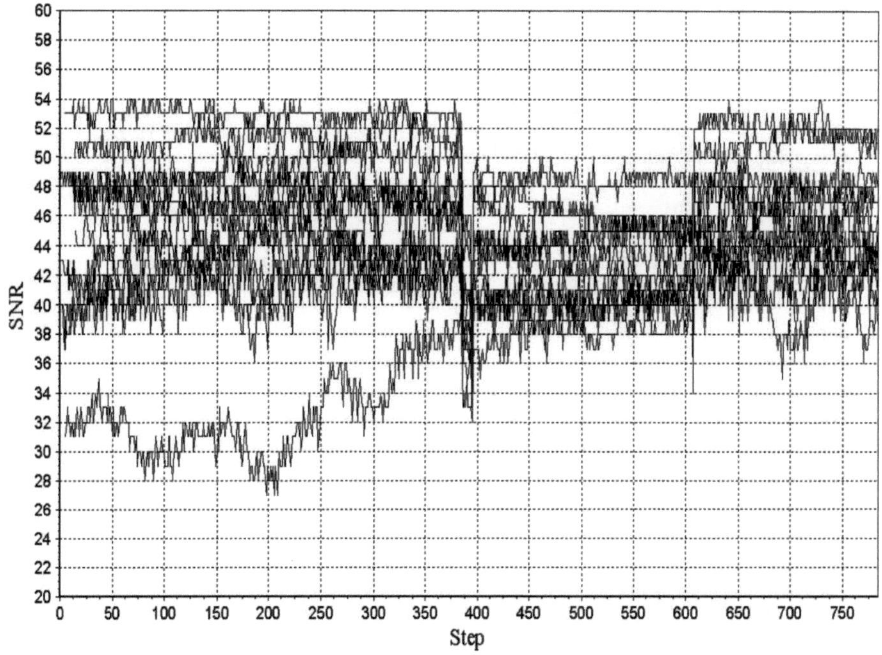

Fig. 3.12 Signal-to-noise ratios for tracked SVs in the LAAS reference receiver in the presence of the INMARSAT signal

Thus, the introduction of new LAAS quality monitoring operations, as well as estimation of the ground MP error variance and variances of noise errors on the LAAS and onboard the aircraft, makes it possible to improve the positioning accuracy and integrity.

The LAAS quality monitoring accompanied by the detection of radiointerference and alerting the AC pilot of these effects supports flight safety.

Figures 3.12 and 3.13 show the results of full-scale LAAS tests in the presence of radiointerference effects (Fig. 3.12 shows the signal-to-noise ratio of the LAAS receiver with the INMARSAT signal effect, and Fig. 3.13 shows the current GISR value for the LAAS).

During the tests, the international communication system INMARSAT operating in the frequency range of 1626.5–1660 MHz was operating near the LAAS antenna.

As a result, the level of the signal-to-noise ratio in the GPS reception channels was sharply reduced (Fig. 3.8, step: 400–600).

The time dependence of the current GISR values shown in Fig. 3.9 demonstrates the detectability of radio effects on the LAAS.

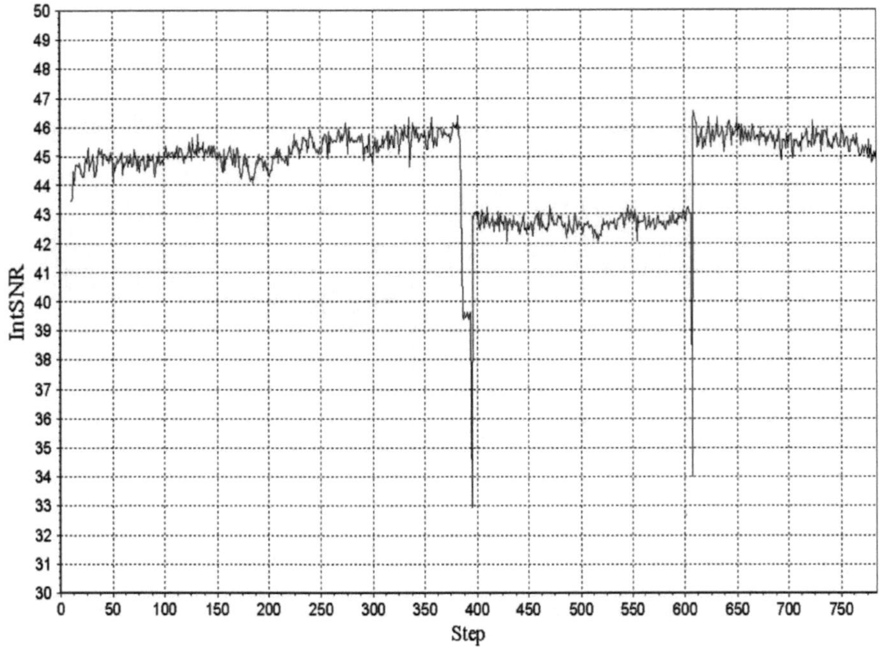

Fig. 3.13 Current GISR value for the LAAS

As an example illustrating the generation of the LAAS GPE variance in accordance with the method described above, Fig. 3.14 shows the results of the PD differential correction for one GPS satellite over a three-hour time interval, and Fig. 3.15 shows the LAAS GAD-A and the root-mean-square value of the ground pseudorange error (GPE RMS). Effects of the mirror MP (with a correlation interval of up to 500 s) appeared at the last hour of the experiment (Fig. 3.14, step: 6500–10,500). Accordingly, the ground PR error during the first two hours of the experiment (step: 0–6500) was formed without measurements of the signal-to-noise ratio of the ground receiver (statistics $\Delta_2^*(k)$ was used). During the last hour of the experiment (step: 6500–10,500), the ground PR error was formed using PR code and phase measurements, as well as measurements of the signal-to-noise ratio of the ground receiver per (3.27).

Let us consider further the method for increasing accuracy and integrity of navigation information for the SLS based on the compensation of errors in phase measurements.

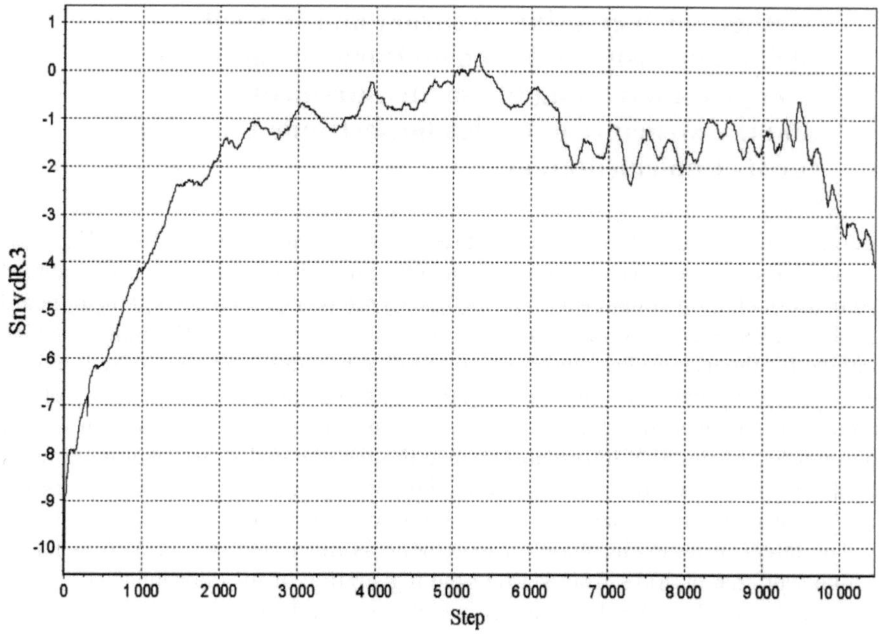

Fig. 3.14 Change of the pseudorange differential correction over time

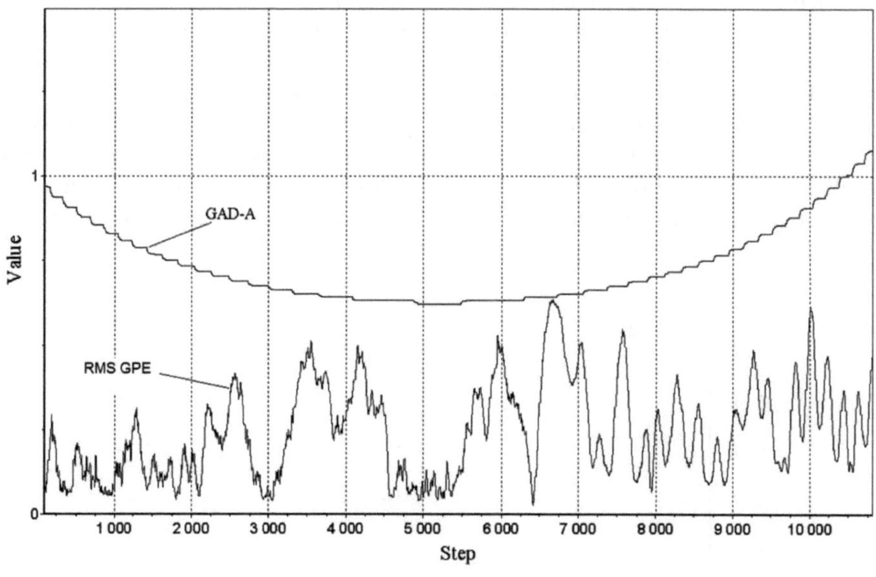

Fig. 3.15 Change of the root-mean-square value of the ground pseudorange error over time

3.3 Method for Increasing Accuracy and Integrity of Guidance Signals Based on Pseudorange Error Compensation Using Phase Measurements and the Structure of the Radioelectronic Complex for Its Implementation

A common drawback of the known methods and devices used in the SLS [17, 36] is that there is no detection and no correction of the jumps in the phase measurements of the phase-lock loop system (PLL) of the LAAS reference receivers and the onboard GNSS/LAAS receiver. At the same time, the requirements for LAAS equipment [20] provide for the use of phase information to refine pseudoranges, but these jumps in phase measurements can introduce significant errors in the measurements of the aircraft coordinates. In addition, these jumps in phase measurements can reduce the integrity of the SLS information, since the recommended integrity monitoring algorithm [16] based on the variance of the coordinate fluctuation error, does not take into account the presence of jumps in phase measurements.

The effect of jumps in phase measurements is illustrated in Fig. 3.16, where it can be seen that at 200 s from starting the registration of the coordinates (*X*, *Y*, *Z*) calculated in the LAAS reference receiver, there was a PLL cycle and, as a result, a jump in the coordinates.

The purpose of the proposed error compensation method is to eliminate such jumps and thereby increase the accuracy and integrity of the SLS navigation information.

A number of works [37–40] are known in which such phase jumps in the satellite radio navigation system are corrected. However, they all have various disadvantages. For example, in [38], the phase jump correction is based on the use of WAAS and EGNOS wide-area systems, which are available only in the USA and Western Europe. In particular, this method cannot be used in polar regions because of the small elevation angle of geostationary SVs. Altmayer [39] considers the detection of phase jumps in a system that uses three sets of the receiver–antenna equipment.

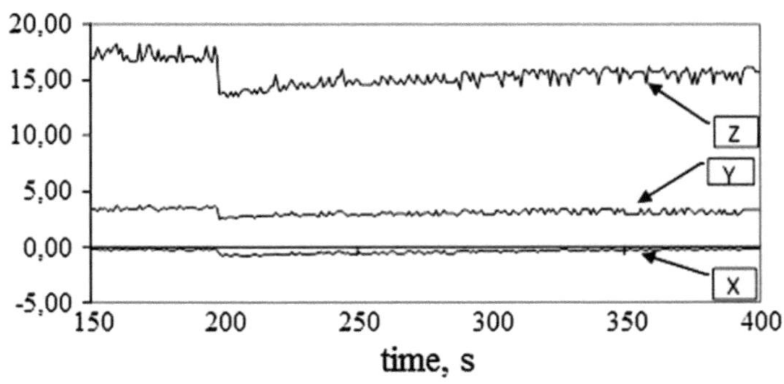

Fig. 3.16 Relative coordinates of the LAAS reference receiver

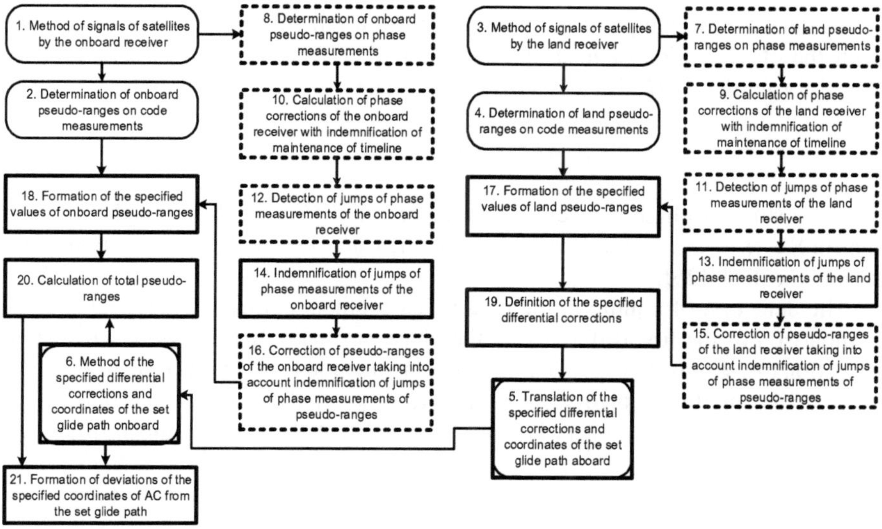

Fig. 3.17 Method for compensation of errors in phase measurements

The method described in [37, 41] is based on the detection of single anomalies. A method for correcting phase jumps in a real-time kinematic system (RTK) using an inertial navigation system was proposed in Ref. [40].

The main disadvantage of the known methods is the low positioning accuracy. One way to improve the accuracy is to use not only the code component, but also the phase component of the radio signal. At the same time, the known methods [16, 21, 41] of joint code-phase filtration provide insufficient accuracy due to the effect of jumps in the PR phase measurement. The task of the proposed method is to increase the accuracy of calculating the aircraft coordinates and the reliability of their determination due to the PR phase measurements taken into account with detection and correction of jumps.

Figure 3.17 shows the sequence of operations in the implementation of the proposed method to compensate for errors in phase measurements. New operations that implement the proposed compensation method are shown in Fig. 3.13 by a rectangular continuous contour and described in detail in [8, 9].

Modern receivers, as a rule, are made with outputs of phase measurements [42]. Pseudoranges in the LAAS reference receivers and onboard GNSS/LAAS equipment are determined by phase measurements. After the determination of the pseudoranges by phase measurements, phase corrections are calculated with compensation of the receivers' clock offset by subtracting the mean value of the phase corrections considered, for example, in [41].

For the purpose of detecting jumps in phase measurements, a well-known algorithm can be used that is based on the detection of jumps in phase measurements in the "sliding window" on the time scale proposed in [37, 41].

To detect these jumps, it is proposed to use a fail-safe nonlinear Kalman filter with the classification of measurements [43]. For this, the first difference of the phase correction $dy_i(k)$ is calculated at the kth time step for the ith SV:

$$dy_i(k) = y_i(k) - y_i(k-1), \tag{3.31}$$

where $y_i(k)$ is a phase correction defined as the difference between the range to the ith SV ($i = 1 \ldots N$) and the PR phase measurement with compensation of the receiver's clock [25].

The detection of a jump in the phase measurements of each SV is reduced to a comparison of the first difference in the phase correction $dy_i(k)$ with two varying detection thresholds $\pm\psi_{\text{nop},i}(k)$ determined by method [43]. In this case, if the value $dy_i(k)$ falls into the region of $\Omega_i \in [-\psi_{\text{nop},i}, +\psi_{\text{nop},i}]$, then a decision is made about the absence of a jump, otherwise, it is decided that there is a jump in the phase measurement.

After making a decision on the presence or absence of a jump in the phase measurement, the parameters of the Kalman filter are calculated using the procedure [43] with the measurement classification.

The detected jumps in phase measurements in the LAAS reference receivers and onboard GNSS/LAAS equipment are compensated by calculating the total correction $\vartheta_i(k)$ for each ith SV that is used to refine the current phase measurement $Y_i(k)$ in accordance with formulas:

$$\vartheta_i(k) = \begin{cases} \vartheta_i(k-1), \text{ if } dy_i(k) \in \Omega_i \\ \vartheta_i(k-1) + dy_i(k) - \frac{1}{N}\sum\limits_{j=1}^{N} dy_j(k), \text{ if } dy_i(k) \notin \Omega_i \end{cases} \tag{3.32}$$

$$Yf_i(k) = Y_i(k) + \vartheta_i(k), \tag{3.33}$$

where $Yf_i(k)$ is the compensated phase measurement of the ith SV.

The initial value of the total correction $\vartheta_i(0) = 0$ changes after the appearance of the jump by Formula (3.32); it is memorized in the variable $\vartheta_i(k)$ and shifts the value of the current phase measurement $Y_i(k)$ in accordance with (3.33).

Then, the PR values of the ground and onboard receivers are corrected by calculating the smoothing correction $\Delta x_i(k)$ of the ith PR [20]:

$$\Delta x_i(k) = (1 - B)[Z_i(k-1) + \Delta Y_i(k)], \tag{3.34}$$

where k is discrete (sampled) time, B is the weighting factor, $Z_i(k-1)$ is the refined value of the ith PR at step $k-1$, taking into account the code and phase measurements, $\Delta Y_i(k) = Y_i(k) - Y_i(k-1)$ is the increment of the ith phase measurements $Y_i(k)$.

According to the PR code measurements $X_i(k)$, the pseudorange value for ground and onboard receivers is refined (updated):

$$Z_i(k) = BX_i(k) + \Delta x_i(k). \tag{3.35}$$

The refined values of the pseudorange for the LAAS ground receiver are used to calculate the refined differential corrections (using the technique of Ref. [16]) and are transmitted to the aircraft in the common information packet together with the coordinates of the intended glide path (according to RTCA DO-246C [20], Baburov [43]).

Onboard, the refined differential corrections are received and used together with the above-mentioned refined values of the onboard pseudoranges to calculate the resulting pseudoranges and deviations of the refined aircraft coordinates from the intended glide path.

Let us now consider how the use of new proposed operations makes it possible to improve the accuracy and integrity of the calculation of the aircraft coordinates.

For aircraft positioning, it is necessary to provide the so-called navigation solution. To provide such a solution using the least squares method [42], at each iteration step, the following increment of coordinates $\Delta\overleftrightarrow{xyz}$ is estimated:

$$\Delta\overleftrightarrow{xyz} = S\Delta Z, \tag{3.36}$$

where S is the projection matrix, ΔZ is an N-dimensional vector of pseudorange deviations.

The vector of pseudorange deviations ΔZ may contain phase jumps when refined pseudorange values are formed. Studies show that the coefficient of transmitting the jump to the aircraft height coordinate can reach 3.7 [44], i.e., with a jump in the phase measurement, for example, of 1 m, the height measurement error will be 3.7 m. We also note that the algorithm for monitoring the integrity of the landing system [16] does not take into account the errors created by jumps in the phase measurements.

Thus, with a jump in the measured coordinate (height), no alert of the reduced accuracy is generated. On account of the proposed method, jumps in the aircraft coordinates are eliminated and their accuracy is increased.

The SLS structure in which the proposed method is implemented is shown in Fig. 3.18, where the newly introduced blocks are shown by a thinner line.

The introduction of blocks determining and processing the refined differential corrections and their use in the SLS together with known units processing the SV signals, the data link blocks and the introduced blocks calculating the refined pseudorange values of the ground and onboard receivers due to the phase components of the pseudorange measurements with detection and correction of jumps, help implement the proposed method of increasing the accuracy and integrity of the SLS information.

The increase in the accuracy of the positioning using the method described above is illustrated in Table 3.2.

The presented results are obtained by semi-realistic simulation of the navigation solution in the onboard GNSS receiver for eight implementations, each lasting 150 s (standard approach time).

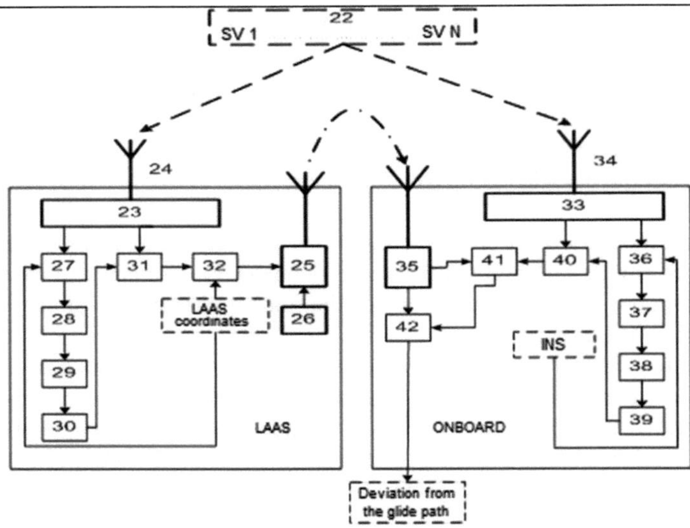

The numbers indicate: 22. Colstellation N of navigation SVs. 23. Ground receiver. 24. Groundfor antenna of satellite signals. 25. Data link transmitter with antenna. 26. Ground database of coordinates of the intended glide path. 27. Ground calculator of phase corrections. 28. Ground detector of jumps in phase measurements. 29. Ground calculator of compensated values of pseudorange phase measurements. 30. Ground calculator of pseudorange smoothing corrections. 31. Ground calculator of refined values of pseudoranges. 32. Calculator of refined differential corrections. 33. Onboard receiver. 34. Onboard antenna of the satellite signal receiver. 35. Data link receiver with antenna. 36. Onboard calculator of phase corrections. 37. Onboard detector of jumps in phase measurements. 38. Onboard calculator of compensated values of pseudorange phase measurements. 39. Onboard calculator for pseudorange smoothing corrections. 40. Onboard calculator of refined values of pseudoranges. 41. Onboard calculator of resulting pseudoranges. 42. Calculator of deviations of the refined coordinates from the intended glide path.

Fig. 3.18 Structural SLS diagram for the implementation of the phase measurement error compensation method

The initial data contained phase measurements in which jumps were present. Two solutions were formed: one without using the method for phase jump correction, and the second one using this method. For each implementation, the root-mean-square error (RMS) was estimated.

The average value of the F_{Acc} factor for all implementations presented in Table 3.2 is 2.9.

Table 3.2 Positioning error estimate

D_{RWY} (km)	1	2	3	4	5	6	7	8
σ_0 (m)	0.3	0.1	0.2	0.4	0.4	0.6	0.8	0.7
σ_1 (m)	1.2	0.6	1.2	1.1	1.4	0.9	1.3	1.5
F_{Acc}	4.0	6.0	6.0	2.7	3.5	1.5	1.6	2.1

Note σ_0 are RMSs calculated by a standard method, σ_1 are RMSs calculated with the method for compensating jumps in phase measurements, $F_{\mathrm{Acc}} = \sigma_0/\sigma_1$

Table 3.3 Values of protection levels

D_{RWY} (km)	0	1	2	3	4	8	16
L (m)	7.2	6.4	6.9	4.3	6.7	4.1	5.5
L^* (m)	4.3	4.1	5.1	3.3	3.7	2.4	3.3
$F_{\text{Integrity}}$ (L)	1.7	1.4	1.1	1.2	1.8	1.9	1.6
V (m)	3.6	4.6	2.9	3.8	4.1	2.8	2.5
V^* (m)	2.1	3.2	1.8	2.1	3.2	1.9	1.9
$F_{\text{Integrity}}$ (V)	1.6	1.4	1.9	1.4	1.5	1.7	1.5

Note Values of protection levels calculated using the method for compensating jumps in phase measurements. $F_{\text{Int}}(L) = L/L^*$, $F_{\text{Int}}(L) = V/V^*$. The average $F_{\text{Int}}(V)$ value is 1.5, the average $F_{\text{Int}}(L)$ value is 2.3

The results of calculating the protection levels (horizontal L and vertical V, (see (3.11), (3.12)) for various distances from the threshold are shown in Table 3.3. The calculations were performed simultaneously with the calculations of the coordinates that were formed without the use of the considered method for compensating jumps in phase measurements, and then with its use.

The analysis of the results of the semi-realistic simulation presented above shows that the use of the proposed method makes it possible to increase the accuracy of the measurements by a factor of 2.9 (see the F_{Acc} factor) and to reduce the calculated value of the horizontal protection level by a factor of 1.9 and the vertical protection level be a factor of 1.6 times, which reduces the probability of false alarms by 2–4 orders of magnitude (see Sect. 3.2) and reduces the probability of the data output continuity loss by the same amount.

Next, consider the method of using pseudosatellites in the SLS construction to improve the accuracy, integrity, and continuity of information.

3.4 Method for Increasing Accuracy, Integrity, Continuity, and Availability of Guidance Signals Based on the Use of Pseudosatellite Signals and the System Structure for Its Implementation

One way to ensure high accuracy, integrity, and continuity of navigation information in the GNSS is common application of the information coming from the GNSS SVs and ground pseudolites (PS) [45]. In general, the concept of "pseudolite" includes a source of a radio navigation signal with a structure similar to the GNSS SV signals but located on the surface of the earth.

The main advantages of using PSs include the capability of forming a navigation signal that does not contain ephemeris, ionospheric and tropospheric errors, as well as a significant increase in the accuracy of determining the altitude from GNSS signals

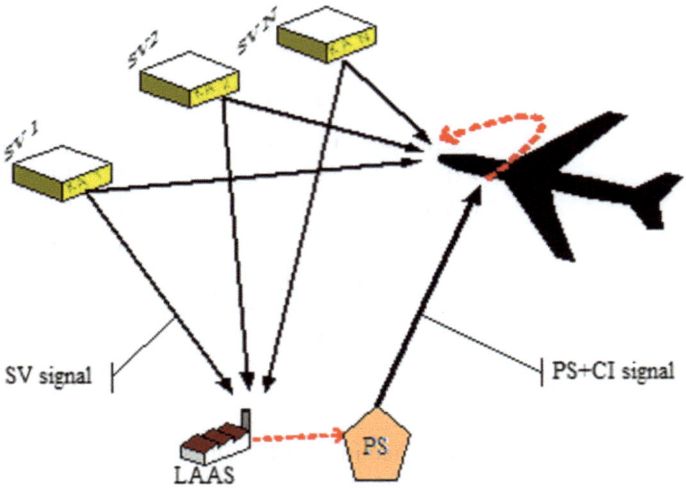

Fig. 3.19 Use of PSs for the construction of an SLS with a channel for transmitting CI through PSs

during flights in the vicinity of the PSs. The PS radio channel can also be used as a channel for transmitting the LAAS corrective information, instead of the LAAS VDB transmitter.

Figure 3.19 is an illustration of the SLS construction when using the LAAS with a channel for transmitting corrective information (CI) over the PS channel.

The purpose of this SLS construction method is to improve accuracy, integrity, and continuity of navigation information. The main difference between this method and the others that also use PSs is that its practical implementation does not require installation of new onboard GNSS/LAAS equipment, and it is possible to use existing onboard GNSS receivers. This is especially important for the upgrade of aircraft that are already in operation, since there are no additional costs for new (non-standard) GNSS receivers.

The currently known systems [17, 46, 47] using PS technologies have various disadvantages including the difference in the carrier frequency of the PS signals from the standard carrier frequency of the GNSS L1 signals. This prevents from the use of standard onboard GNSS receivers on the aircraft, and when using the L1 frequency, the problem of matching the power of the PS and SV signals occurs.

Since the power of the PS signals is always higher than the power of the SV signals, it is necessary to ensure that the levels of these powers are matched within the same order in order to minimize interference caused by PS and SV signals.

In addition, the signals emitted by the PSs are characterized by a larger dy power namic range [48]; and this, when standard SV signals are used for the PSs, results in either a significant reduction in the PS coverage area or a decrease in accuracy.

When receiving satellite signals, the signal-to-noise ratio lies within the range of 30–50 dBHz. If, at the landing decision point, which corresponds to a height of

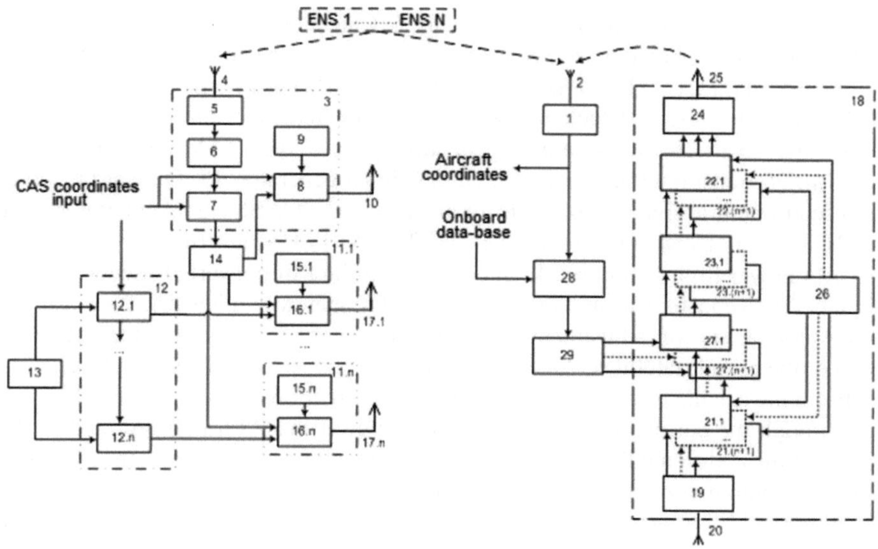

Fig. 3.20 Structural SLS diagram with PSs

60 m in Category I landing systems, the PS signal strength is determined by the signal-to-noise ratio of 50 dBHz, then the maximum slant range to the PS at which the signal-to-noise ratio decreases to 30 dBHz will be only 600 m, while the SLS range must be at least 36 km.

Thus, in order to construct an SLS using PSs, it is necessary to expand the area of reliable reception of the PS signals, which is achieved, for example, by limiting the level of the PS signal radiated power.

In known systems with PSs, such a limitation of the radiated power is made by a parametric method, by forming special directional patterns of the PS transmitting antennas. The construction of such antennas is described, for example, in [48, 49]. However, this method of limiting power is very complex and time-consuming.

The structural SLS diagram implementing the proposed method is described, in detail, in the author's paper [9] and shown in Fig. 3.20. The diagram works as follows.

Modules 1–17 (see Fig. 3.20) perform functions similar to the standard SLS construction diagram.

LAAS reference receivers, PSs, and onboard GNSS/LAAS receiver receive SV radio navigation signals; the PS calculator processes the received signals and generates an array of pseudoranges to the SVs and calculates the PS coordinates.

These data are then fed to the input of the correction calculator that also receives the specified PS coordinates.

The calculator calculates corrections in accordance with the requirements of [16] and generates the CI. The numbers in Fig. 3.20 indicate the following: 1. Onboard receiver for SV signals, 2. Onboard receiving antenna, 3. PSs, 4. Ground receiving antenna, 5. Ground receiver for ENS signals, 6. Calculator of measured parameters, 7. Calculator of corrections, 8. PS transmitter, 9. Reference generator, 10. PS transmitter antenna, 11.1 … 11.*n* PSs, 12. Calculator of APS (additional pseudosatellites) optimal coordinates, 13. Block of information parameters of the airport, 14. Distribution center, 15.1 … 0.15.*n* Reference generators, 16.1 … 16.*n*. ASV transmitters, 17.1 … 17.*n*. PS transmitter antennas, 18. GNSS signal generator, 19. Onboard receiver for pseudosatellite signals, 20. Receiver antenna of the GNSS signal generator, 21.1 … 21. $(n + 1)$ Downconverters, 22.1 … 22. $(n + 1)$ Upconverters, 23.1 … 0.23. $(n + 1)$ Power limiting units, 24. Onboard transmitter of converted pseudosatellite signals, 25. Repeater transmitter antenna, 26. Frequency synthesizer, 27.1 … 27.$n + 1$ Automatic power adjustment units, 28. Calculator of ranges from the aircraft to each of the pseudosatellites, 29. Functional range converter to the signal level.

From the calculator output, the pseudorange corrections to each of the N SVs are fed through the correction distributor to the input of the PS transmitter and to the inputs of the additional PS transmitters. At the same time, signals from the reference generator output come to the drive input of the PS transmitter. The drive inputs of all additional PS transmitters also receive signals from all corresponding reference generators.

Signals from all PSs are received onboard the aircraft by the GNSS signal generator antenna of specially introduced onboard receiver for PS signals and are fed to the inputs of downconverters, which reduce the frequency to a range that allows efficiently limiting the signal power by means of power limiting units. Signals of limited power are increased in frequency by upconverters to the frequency range of the GNSS signals. These signals of increased frequency are transmitted over the onboard transmitter of PS signals through the GNSS signal generator antenna. Thus, the GNSS signal generator provides the information link "PS transmission-reception" with the onboard GNSS/LAAS receiver.

The reduction and increase in the frequency in the GNSS signal generator are controlled with a frequency synthesizer.

The used PSs have frequencies offset relative to the SV signals in order to eliminate mutual interference of the PS and SV signals. For the PS operation, it is advisable to use the frequency navigation range of 960–1215 MHz, since the well-known DME navigational systems, SRNSs, operating in this range use pulse signals with a high-duty cycle that ensures their good compatibility with continuous pseudorandom GNSS signals [16].

The useful effect of the SLS shown in Fig. 3.20 is achieved by limiting the signal power coming from the PSs to the power level of the satellite signals. When the PS range is changed, the dynamic range of the PS signal power change is reduced due to the action of the power limiter, and this increases the range of reliable reception of the PS signals without interference to satellite signals. As a consequence, a universal GNSS/LAAS receiver can be used onboard, and no changes are made to the algorithm of the receiver's secondary processing, or these are minimal.

Let us consider the feasibility of the proposed diagram. At the present time, special microchips have been developed that operate at an intermediate frequency of up to 500 MHz and are designed to limit the signal power in transmitting and receiving devices for digital signals. As an example, consider the power limiting device on the AD8367 microchip (made by Analog Devices), which is an amplifier with a passive attenuator and a power meter with a square-law detector. This construction of the limiter allows for the reduction of nonlinear distortions and the effect of temperature on the device operation. The microchip has an adjustment range of 45 dB and a limitation error of less than 1 dB. If the receiver used has a dynamic range of 20 dB and the limit range for the GNSS signal generator power is 40 dB, the range of power change for the received PS signal will be 60 dB. With a margin of 10 dB to compensate for the effect of the GNSS signal generator receiving antenna pattern in conditions of the aircraft body oscillations, we obtain an allowable change in the field strength of the PS signal equal to 50 dB. The PS signal can be received within the range of 60 m to 19 km. This range is less than the range coverage of the meter-wave landing system (46 km along the azimuth channel, 18.5 km along the glide path).

To extend the range coverage, additional solutions are needed.

For this, before power is limited, automatic power adjustment is performed using automatic power adjustment units whose inputs receive signals from the corresponding outputs of the downconverters, and it is only after this automatic power adjustment that they enter the input of the standard GNSS/LAAS receiver.

Automatic power adjustment is performed as a function of the distance between the aircraft and the PSs. For this purpose, the calculator determines distances from the aircraft to each of the PSs.

These distances are converted by the functional converter into signal levels necessary for use as control signals of automatic power adjustment units.

As an automatic power controller, an amplifier with digital gain control can be used, for example, the AD8369 microchip (made by Analog Devices). Since this amplifier has a logarithmic conversion characteristic, the functional converter shall calculate the dependence $\log\left(\frac{1}{r^2}\right)$, where r is the distance between the aircraft and the PS.

When using an amplifier with digital control (for example, the AD8369 microchip has an adjustment range of 45 dB), the same characteristics can be obtained as with a power limiter; however, there is a danger of overloading the GNSS/LAAS onboard receiver by a strong signal due to the spread of the PS transmitter antenna gain and the GNSS signal generator receiver antenna gain onboard the aircraft. Therefore, the best stabilization of the signal power level when the range is changed can be obtained by using an automatic power controller, the output of which has a series connection with a power limiter. In this case, the range of power control for the received signal exceeds 80 dB, while the range of 0.06–46 km corresponds to a change in the signal power of 58 dB. Thus, with the combined use of the automatic controller and the power limiter, the required range of 46 km is provided with a large margin for changing the gain of the GNSS signal generator receiving antenna onboard the aircraft during maneuvering (about 22 dB).

As frequency converters, it is advisable to use a mixer AD8343 (made by Analog Devices).

In this case, it is sufficient to use a matching gain stage with a bandpass filter as a receiver (for example, with a SAW filter) at the input, which selects the frequency

Table 3.4 Operating frequencies of pseudosatellites

f_{PS} (MHz)	f_1 (MHz)	f_2 (MHz)	f_{ADJ} (MHz)
1100	500	1000	600
1050	550	1100	500
1000	600	1200	400
950	650	1300	300
900	700	1400	200

range of the PS signals. And, it is sufficient to use a matching gain stage with a bandpass filter as a transmitter at the output that limits the emission of signals outside the GNSS band. Frequency converters require local oscillator voltages that are generated in the frequency synthesizer.

Consider the requirements for the frequency synthesizer used in the SLS with the PSs shown in Fig. 3.20.

The frequency conversion algorithm is described by the following relations:

$$F_{ADJ} = f_{PS} - f_1; \; f_{L1} = f_{ADJ} + f_2; \; f_{ADJ} < f_{LIM}, \tag{3.37}$$

where F_{ADJ} is the frequency at which the PS signal power is adjusted and limited, and its value should not exceed the limiting value f_{LIM} determined by the frequency range of the limitation and automatic power control units; f_{PS} is the frequency of the PS signal; f_{L1} is the frequency of the SV signal in the L1 band; f_1 and f_2 are the frequencies of the local oscillator voltages that are generated in the frequency synthesizer for each PS.

In addition, the f_{PS} values should be in the frequency range of 960–1215 MHz, in which the well-known DME navigation systems, SRNSs, operate. To simplify the implementation of the frequency synthesizer, it is also advisable that the condition: $f_2 = mf_1$ is met, where m is an integer.

For a given f_{PS} value, the following relationship is satisfied:

$$(m - 1) \cdot f_1 = f_{L1} - f_{PS} = \text{const.}$$

With increasing m, f_1 decreases, which leads to an increase in $f_{ADJ} = f_{PS} - f_1$ value and violation of the condition $f_{ADJ} < f_{LIM}$. For this reason, it is advisable to take $m = 2$.

The procedure for determining the frequencies of the local oscillator voltages in the frequency synthesizer is as follows:

- the f_{PS} value is specified,
- values $f_1 = f_{L1} - f_{PS}$ and $f_{ADJ} = f_{PS} - f_1$ are calculated,
- the condition $f_{ADJ} < f_{LIM}$ is verified.

An example of calculations (for $f_{L1} = 1600$ MHz) is given in Table 3.4.

Note that the condition $f_{\text{ADJ}} < f_{\text{LIM}}$ places an upper limit on the f_{PS} value. If $f_{\text{LIM}} = 500$ MHz, then from the condition $f_{\text{L1}} - 2f_1 = f_{\text{ADJ}} = 500$ MHz at $f_{\text{L1}} = 1600$ MHz, we find $f_1 = 550$ MHz, and from the condition $f_{\text{PS}} - f_1 = f_{\text{ADJ}} = 500$ MHz, we find $f_{\text{PS}} = 1050$ MHz. Consequently, PSs can operate in the first (lower) frequency band of the DME beacon (960–1025 MHz) and in the entire frequency range of the SRNS DME (939.6–1000.5 MHz).

Such requirements for a frequency synthesizer in a pseudosatellite GNSS signal generator are met, for example, by the ADF4360-7 microchip (made by Analog Devices) with a phase-lock loop system in the synthesizer and a built-in controlled generator. A similar principle of constructing a frequency synthesizer makes it possible to reduce the level of combination frequencies in the spectrum of the converted PS signals [50]. The microchip has a main signal output with a frequency in the range of 350–1800 MHz (f_2) and an additional signal output with half frequency in the range of 175–900 MHz (f_1).

The complexity and, consequently, the cost of manufacturing the proposed system on the above-mentioned element base is much lower than the development and manufacture of a new GNSS/LAAS receiver and PSs.

Mathematical modeling of a landing system based on GLONASS/GPS satellite navigation systems augmented by one or several PS [43], in which the proposed landing system variants can be used, show that the accuracy of positioning and, accordingly, the guidance signals, increases by a factor of 1.4. At the same time, the probability of the navigation solution under reduced visibility of the SV constellation increases by 25–50%. Obviously, the continuity of the guidance signals generation will increase by the same amount.

3.5 Conclusions

This chapter introduces new methods and suggests diagrams for constructing devices to improve flight safety. These methods are based on the laws of radio wave propagation and provide increased accuracy, integrity, and continuity of navigation information generated in ground systems of GNSS augmentations (LAAS), onboard GNSS/LAAS equipment, and onboard devices for pseudosatellite signal reception.

This chapter describes a method for constructing and using diagrams of the volumetric distribution of errors in multipath propagation of radio waves that are defined on the basis of statistical accumulation of information on the characteristics of the received signals on the LAAS and known scattering diagrams depending on the aircraft structural elements, which makes it possible to increase the accuracy of estimating the LAAS differential corrections by a factor of 1.5–6, 0. At the same time, the probability of false alarm is reduced by several orders of magnitude, the continuity of the guidance data is increased, and as a result, the integrity of the data and flight safety is increased.

It also describes a method for compensating pseudorange errors in the ground and onboard GNSS subsystems through the identification and elimination of jumps in phase measurements, which makes it possible to increase the accuracy of determining the coordinates by a factor of 2 or more. Important is the fact that these errors are not taken into account by standard integrity monitoring algorithms.

The chapter describes as well a method for ensuring the integrity and continuity of information based on the use of an integrated signal-to-noise ratio, which allows maintaining the specified accuracy of coordinate information in challenging interference environment. This method helps detect the appearance of interference and thereby increase the integrity of information.

Also a method is described that involves pseudosatellites in the SLS and makes it possible to overcome the main problem of widespread introduction of pseudosatellites, that is, a large dynamic range of signal level changes at the input of the onboard receiver, and to maintain the standard structure of the onboard equipment complex. It also increases the accuracy of determining the coordinates and generated guidance signals two- or threefold depending on the navigation constellation used; integrity of navigation information is doubled; and continuity of navigation information can increase manyfold.

In the next chapter, we will consider methods and devices to improve flight safety, the use of which is reasonable in collision avoidance systems (CASs).

References

1. Bellman R, Zadeh L (1976) Decision-making under vague conditions. In: Analysis issues and decision-making procedures. Mir, Moscow, pp 172–215 (in Russian)
2. Sauta OI (2011) Use of satellite navigation information to improve the efficiency of armament. In: Interuniversity scientific and technical seminar. Abstracts of the report. Mikhailov Military Artillery Academy (in Russian)
3. Sauta OI (2010) Recommendations on the integrated use of satellite technologies in control systems (in Russian)
4. Baburov VI, Volchok JG, Galperin TB, Gubkin SV, Dolzhenkov NN, Zavalishin OI, Kupchinsky EB, Kushelman VY, Sauta OI, Sokolov AI, Jurchenko JS (2009) Airplane landing method using a satellite navigation system and a landing system based thereon. WO 2009/011611 A1, publication date 22.01.2009
5. Baburov VI, Volchok JG, Galperin TB, Gubkin SV, Dolzhenkov NN, Zavalishin OI, Kupchinsky EB, Kushelman VY, Sauta OI, Sokolov AI, Jurchenko JS (2008) Airplane landing method using a satellite navigation system and a landing system based thereon. Patent No. RU 2 331 901, publ. 20.08.2008, Bul. No. 23 (in Russian)
6. Baburov VI, Volchok YuG, Galperin TB, Gubkin SV, Sauta OI, Sokolov AI, Jurchenko YuS (2010) Airplane landing method using a satellite navigation system and a landing system based thereon. Patent No. RU 2 385 469, publ. 27.03.2010, Bul. № 9 (in Russian)
7. Baburov VI, Volchok YuG, Galperin TB, Gubkin SV, Sauta OI, Sokolov AI, Chistyakova SS, Jurchenko YuS (2009) Airplane landing method using a satellite navigation system and a landing system based thereon. Patent No. RU 2 371 737, publ. 27.10.2009, Bul. № 30 (in Russian)
8. Baburov VI, Volchok YuG, Galperin TB, Gubkin SV, Ivantsevich NV, Sauta OI, Sokolov AI, Chistyakova SS, Jurchenko YuS (2012) Satellite radio navigation landing system using pseudosatellites. Patent No. RU 2 439 617, publ. 10.01.2012, Bul. № 1 (in Russian)

9. Baburov VI, Galperin TB, Gerchikov AG, Ivantsevich NV, Sayuta OI, Sokolov AI, Chistyakova SS, Jurchenko YuS (2012) Complex method of aircraft navigation. Application No. 2012136399, priority of 17.08.2012 (in Russian)
10. Sauta OI (1986) Obtaining estimates of statistical characteristics of SRNS measurements on the approach trajectories. ChTP, State reg. No. I-85628 (in Russian)
11. Sauta OI, Kurochkina SL (1989) Algorithm for analyzing the quality and increasing the reliability of information using a filter with several radio technical meters. ChTP, State reg. No. G-07757 (in Russian)
12. Ivanov YuP, Nikitin VG, Rogova AA, Sauta OI, Sobolev SP (2006) Analysis of the integrity of the satellite navigation landing system. In: NSTU Collection of scientific papers, Issue 2(44), 188 p, pp 9–20. NSTU Publ. House, Novosibirsk (in Russian)
13. Sauta OI (2006) Method for assessing the integrity of the satellite navigation system. In: Ivanov YuP, Nikitin VG, Rogova AA, Sobolev SP (eds) SPbGETU "LETI" bulletin. Series "Radiolocation and radio navigation", no 5, pp 69–77 (in Russian)
14. Sauta OI, Sokolov AI, Yurchenko YuS (2009) Estimation of data generation accuracy in the local area augmentation system in aviation. In: Radioelectronics issues. Series RLT-2009-Issue 2, pp 183–193 (in Russian)
15. US Patent No. 5 361 212, cl. G01S5/00 application of 11.02.92., publ. on 11/01/1994
16. Annex 10 to the Convention on International Civil Aviation (2006) Aeronautical telecommunications. Radio Navigation Aids. 6th edn, vol 1. (in Russian)
17. Dmitriev PP et al (1993) Network satellite radio navigation systems. In: Shebshayevich VS (edn) 2nd edn. Radio and Communication, Moscow, 408 p (in Russian)
18. Patent of the Russian Federation No. 2 237 256, cl. G01S5/00, H04B1/06, application of February 21, 2001, publ. on 27 Sept 2004 (in Russian)
19. Solovyev YuA (2003) Satellite navigation and its applications. Eco-Trends, Moscow, 326 p (in Russian)
20. RTCA DO-246C (2005) GNSS based precision approach local area augmentation system (LAAS)—signal-in-space interface control document (ICD) [Electronic resource]. Radio Technical Commission for Aeronautics
21. Sokolov AA (2005) Estimation of the LAAS differential correction errors in aviation. Abstract of thesis … and Tech Sciences. S.-Pb (in Russian)
22. Sokolov AI (2006) Monitoring phase measurements in reference receivers of ground local segmentation systems GPS/Glonass. In: Sokolov AI, Chistyakova SS (eds) Publication House SPbGETU "LETI"/SPbGETU "LETI", 2006.-Issue 2. Radioelectronics and Telecommunications, pp 27–32 (in Russian)
23. Sleewaegen J (1997) Multipath mitigation, benefits from using the signal-to-noise ratio. In: ION GPS-1997: proceedings of the 10th international technical meeting of the satellite division of the institute of navigation. Kansas City, pp 531–541
24. Kazarinov YuM (1985) Designing of radio signal filtering devices. In: Kazarinov YuM, Sokolov AI, Yurchenko YuS (eds) under the general editorship by Kazarinov YuS. LETI, Leningrad, 160 p (in Russian)
25. Annual report of the Interstate Aviation Committee (IAC) "Flight Safety Status in 2011" [Electronic resource], www.mak.ru (in Russian)
26. Chistyakova SS (2009) Research and development of methods for detection and correction of jumps in phase measurements in the GNSS-based instrument landing system using. Thesis of technical sciences, SPbGTU "LETI", St. Petersburg (in Russian)
27. Terrain Awareness And Warning System (TAWS). ARINC 763, Recommendations dated 10 Dec 1999 (in Russian)
28. Akos D (2000) Development and testing of the stanford LAAS ground facility prototype. In: Akos D, Gleason S, Enge P, Luo M, Pervan B, Pullen S, Xie G, Yang J, Zhang J (eds) NTM 2000: Proceedings of 2000 National technical meeting of the institute of navigation. Anaheim, 2000, pp 210–219
29. Patent of the Russian Federation No. 2331901, cl. G01S5/02, H04B1/06, application No. 2007128023/09 (030512) of 17.07.2007, positive decision dated 5 Feb 2008, publ. BIZ No. 23 on 20 Aug 2008 (in Russian)

30. Category I Local Area Augmentation System Ground Facility [Electronic resource]: Specification FAA-E-2937A. U.S. Department of Transportation Federal Aviation Administration, 17 Apr 2002. Access mode: http://gps.faa.gov/Library
31. Jan S-S (2003) Aircraft landing using a modernized global positioning system and the wide area augmentation system. A dissertation submitted to the department of aeronautics and astronautics and committee on graduate studies of Stanford University in partial fulfillment of the requirements for the degree of doctor of philosophy. May 2003. [Electronic resource]. Access mode: http://waas.stanford.edu/pubs/index.htm
32. Patent of the Russian Federation No. 2 331 901, cl. G01S5/02, H04B1/06, publ. BIZ No. 23 on 20 Aug 2008 (in Russian)
33. Dryagin DM (2006) Complex enhanced ground proximity warning system with enhanced functionality and software algorithms that minimize the false alarm probability [Electronic resource]: Thesis and technical sciences: 05.11.03. RSL, Moscow (in Russian)
34. Zaezdny AM (1969) Basics of calculations for statistical radioengineering. Moscow Svyaz (in Russian)
35. Kazarinov YuM, Sokolov AI, Yurchenko YuS (1985) Design of radio signal filtering devices. LSU Publ. House, Leningrad (in Russian)
36. US Patent No. 6 469 663 B1, cl. G01S 5/02, application of 24.10.2000, publ. on 22.10.2002
37. Chistyakova SS (2007) Study of the effect of phase measurements jumps on errors in determining the aircraft coordinates in the instrument differential satellite-based landing system. In: 62nd scientific and technical conference on the radio day. Proceedings of the conference. Apr 2007. St. Petersburg. Publ. House SPBGETU "LETI" (in Russian)
38. US Patent No. 6 166 683, cl. G01S 5/02, application of 19.12.1998, publ. on 26.12.2000
39. Altmayer C (2000) Cycle detection and correction by means of integrated system. ION NMT 2000, Anaheim, CA
40. Normark P et al (2001) The next generation integrity monitor testbed (IMT) for ground system development and validation testing presented Sep 2001 at the Institute of Navigation's GPS Conference, Salt Lake City, UT
41. Grishin YuP, Kazarinov YuM (1985) Fail-safe dynamic systems. Radio and Communication, Moscow, 176 p (in Russian)
42. Global Navigation Satellite System (2008) GLONASS. Interface control document. Version 5.1. RNII of Space Instrumentation, Moscow (in Russian)
43. Baburov VI, Vasilyeva NV, Ivantsevich NV, Panov EA (2005) Joint use of the fields of satellite radio navigation systems and pseudo-satellite networks. S.-Pb., Publ. House "Agency RDK-Print", 264 p (in Russian)
44. Sauta OI, Sauta AO, Chistyakova SS, Yurchenko YuS, Sokolov AI, Sharypov AA (2011) Effect of the multipath signal propagation on measurement errors in the global navigation satellite system. In: Fourth all-Russian conference "fundamental and applied position, navigation and time support" (PNT-2011). Theses of reports. Institute of Applied Astronomy, Russian Academy of Sciences, St. Petersburg, pp 224–226 (in Russian)
45. GPS Precision Approach And Landing System For Aircraft. Patent Number: 5 311 194, Date of Patent: 10 May 1994
46. Satellite approach and landing radionavigation system. RF Patent 2236020 GO1S 5/02, application of 19.09.2002, publication on 10.09.2004 (in Russian)
47. Martin S (1999) Antenna diagram shaping for pseudolite transmitter antennas—a solution to the near-far problem. ION GPS'99, p 1479
48. Bartone C, van Grass F (1997) Airport pseudolite for precision approach applications. ION GPS'97, p 1841
49. Alekseev OV, Golovkov AA, Mitrofanov AV, Polevoy VV, Solovyev AA (2003) High- and ultrahigh-frequency generators: study guide for universities. Higher School, Moscow (in Russian)
50. Baburov VI, Volchok JG, Galperin TB, Gubkin SV, Maslov AV, Sauta OI (2007) Method for preventing an aircraft collision with a terrain relief and a device based thereon. WO 2007/123438 A1, publication date 01.11.2007

Chapter 4
Methods for Improving Flight Efficiency and Safety Based on Technologies Applicable in Collision Avoidance Systems

This chapter describes methods, algorithms, and devices, the use of which in collision avoidance systems (CAS) improves flight safety, especially at low altitudes in the terminal area, and under the conditions of hazardous terrain. The proposed methods implement those most relevant areas of enhancing flight safety that have been reviewed in Chap. 2. Detailed descriptions of the most considered methods are given in [1–10].

The proposed methods make it easier for the pilot to make the decision on certain maneuvers aimed to avoid terrain and artificial obstacles and, more importantly, make it possible to prevent the aircraft from encountering potentially hazardous situations.

Particular attention is paid to the methods of providing information on displays to inform the pilot about the current and predicted aircraft position. All proposed methods are based on the use of GNSS technologies, including the use of augmentation systems.

4.1 Method for Improving Flight Safety by Generating a Warning About a Potential Collision Based on the Three-Dimensional Synthesis of the Underlying Surface Sections and the Display of Hazardous Elements

Figure 4.1 illustrates the principle of the "forward terrain assessment" function in the CAS. The point of this mode is that in the direction of flight (ahead), a protective space is synthesized, the shape and size of which (depth, width, length) depend on the current parameters of the aircraft flight (velocity vector, flight phase, track angle rate of change, etc.). At present, a number of methods for preventing collisions are known [11–26], the common disadvantage of which is that they do not provide the pilot with complete information about the nature of the forecasted hazard and, as a result, require additional time for making a decision on the necessary maneuvers to avoid collisions

© Springer Nature Singapore Pte Ltd. 2020
Baburov S.V. et al., *Development of Navigation Technology for Flight Safety*,
Springer Aerospace Technology, https://doi.org/10.1007/978-981-13-8375-5_4

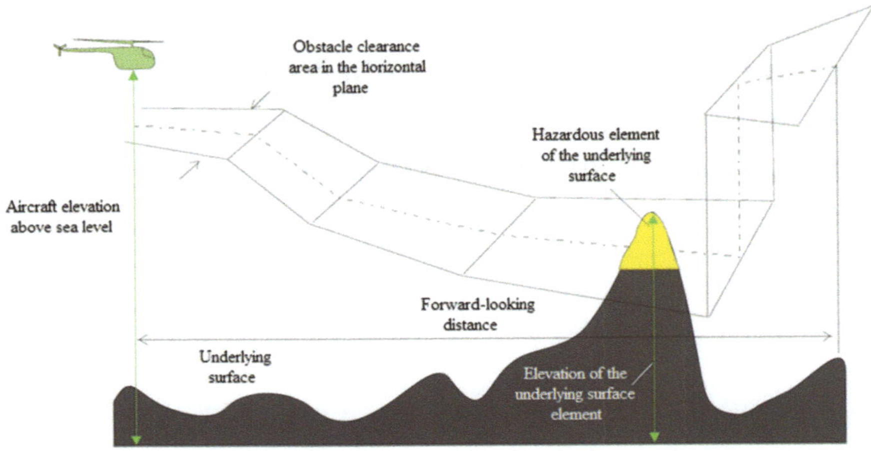

Fig. 4.1 CAS "forward terrain assessment" mode

with the terrain or an artificial obstacle. The idea of the proposed new method consists in the generation and displaying to the pilot of as exhaustive information as possible regarding the nature of the threat by synthesizing three projections of the database for terrain relief and artificial obstacles. These projections are synthesized within the respective information scanning regions formed on the basis of the GNSS information on the aircraft navigation parameters.

The proposed method makes it possible to abandon the synthesis of a three-dimensional representation of the underlying surface, the use of which is difficult for the pilot due to the inevitable distortion of scales, and also requires considerable computational resources for practical implementation.

Consider the content of the proposed method. To simplify the presentation in Fig. 4.2, the main elements are identified on three projections as discussed below. This method is presented in detail in [2–4].

In the proposed method, the aircraft position is determined with the help of the GNSS (see 1 in Fig. 4.2), the parameters of the current aircraft dynamic state are calculated (ground speed W_{gnd}, vertical velocity W_y, track angle TA, turn rate ω_y, etc.) by which the aircraft position is extrapolated over a given time interval t_{pred}. Further, with the current track angle TA, the predicted path is calculated, as well as the protection space (PrSp) associated with the predicted path as shown in Fig. 4.2 on three projections A, B, and C.

In Fig. 4.2, the PrSp is a space domain related directly to the current aircraft position and determined by the shape and orientation in space by the accuracy of determining the coordinates Δ_{coor}, movement parameters (W_{gnd}, W_y, TA, ω_y), the minimum allowable height H_{ma} depending on the flight phase (cruise flight, flight in the terminal area, approach) and the flight mode (level flight, descent, climb).

The PrSp is constructed from the aircraft in the direction of the flight and is intended to generate collision risk alerts for the crew regarding the terrain located in

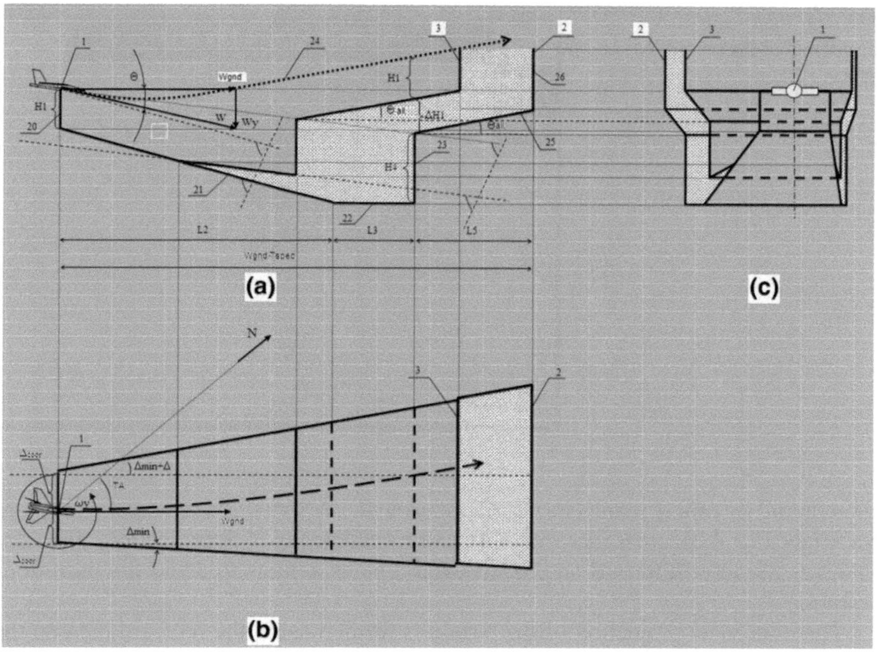

Fig. 4.2 Projections of the protection space: **a** cross projection, **b** plan-view projection, **c** front projection

the area of the intended aircraft position in the nearest given time interval of the flight T_{spec}. Following the results of comparison of the PrSp boundary (see 2 in Fig. 4.2) with the terrain relief database (TDB), an alert is generated on the hazard posed by the terrain ahead of the aircraft. Within the PrSp described above, based on the calculated data, another space similar in form is created—an emergency warning region, where a warning is generated upon crossing its boundary surface (see 3 in Fig. 4.2). Surface 3 is constructed from the calculation of the emergency warning formation in situations where the terrain is at a minimum allowable distance from the aircraft moving in its direction. The calculation for the configuration of protection surfaces 2 and 3 resumes at the required rate and, thus, the PrSp is adapted to the current aircraft dynamic state and to the flight phase and mode.

Simultaneously, a video image of the terrain is formed. The terrain video image is produced on three projections agreed in scale. In addition, all terrain projections are presented on the display in one scale together with the corresponding projections of the protection space, thus forming the combined projections. This method significantly facilitates the crew's decision-making in the current situation, since it makes it possible to assess the degree of proximity to the terrain not only after the alarm is generated, but also before its generation, in the process of checking the terrain display. Thus, it is possible to avoid a hazardous situation onboard the aircraft (alarm

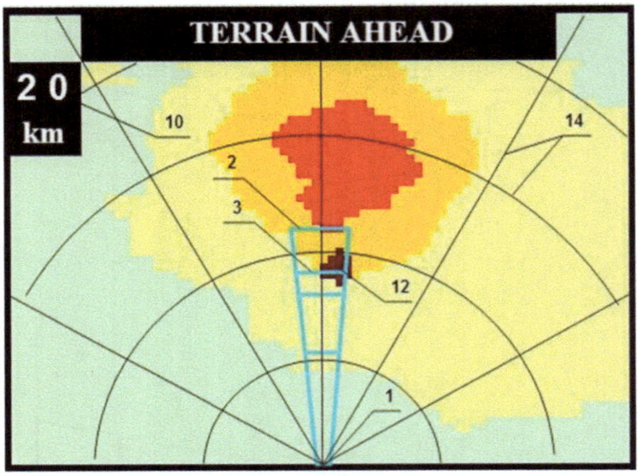

Fig. 4.3 Plan-view projection of the protection space

generation) by timely maneuvering. Comparing the PrSp with the terrain on the plan-view (see Fig. 4.3), cross (see Fig. 4.4), and front (see Fig. 4.5) projections, the crew assesses the terrain hazard and, if necessary, plans actions to avoid the hazardous terrain.

The plan-view projection of the terrain is formed using the onboard data on the terrain, aircraft position, speed, and the track angle. The displayed terrain is painted black, green, yellow, and red, depending on the aircraft height over the terrain (see Fig. 4.6).

This diversity of colors serves as a means of displaying the terrain on the plan-view projection and also (due to the assignment of certain colors to certain elevation ranges) as a means of conveying the degree of the terrain hazard for the aircraft in the formation of any of its three projections.

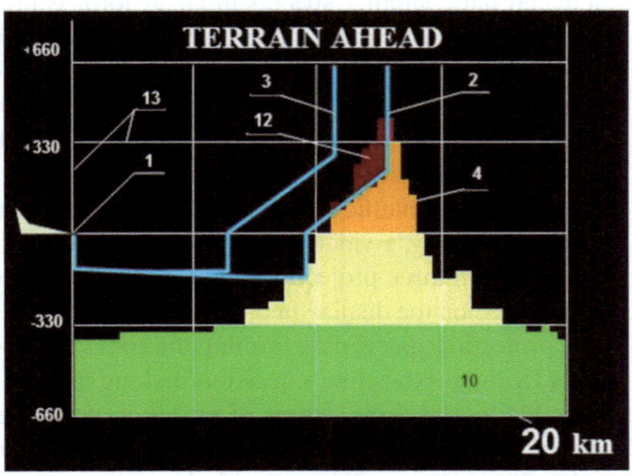

Fig. 4.4 Cross projection of the protection space

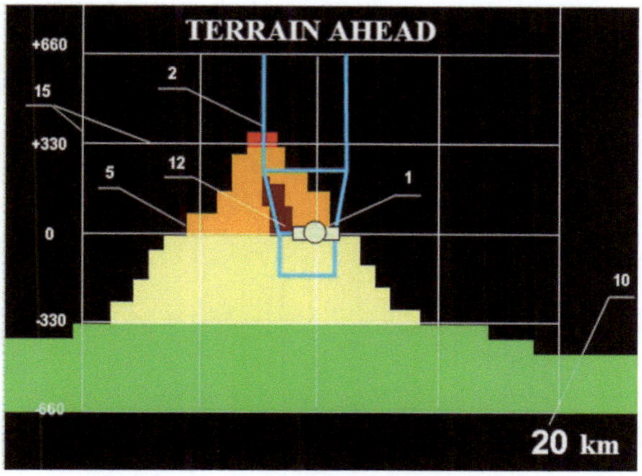

Fig. 4.5 Front projection of the protection space

Unlike the known methods, the proposed method with these colors takes into account not only the terrain elevation on all projections, but also the vertical speed of the aircraft: In the case of the same elevation, the terrain is shown in more alarming colors when the aircraft is descending, and in less alarming colors when the aircraft is climbing as compared to the level flight. Such a shift in the color of the elevation ranges depending on the vertical speed of the aircraft (extrapolation of the video display) is ensured by matching the colors not with a relative elevation of the current aircraft height above the terrain, but with the elevation of the predicted height at which the aircraft will be within the flight time equal to the time t_{pred}. So, in Fig. 4.6 the same color of the terrain corresponds to three different elevations (ΔH_a, ΔH_b and ΔH_c) and vertical speeds ($W_{ya} < 0$, $W_{yb} = 0$ and $W_{yc} > 0$) of the aircraft. Despite the different conditions of the aircraft movement relative to the terrain, it is colored in all three situations ("a," "b," and "c") in the same way, since for all three situations the predicted aircraft elevation (the elevation within the flight time equal to t_{pred}) for the example shown in Fig. 4.6, is the same.

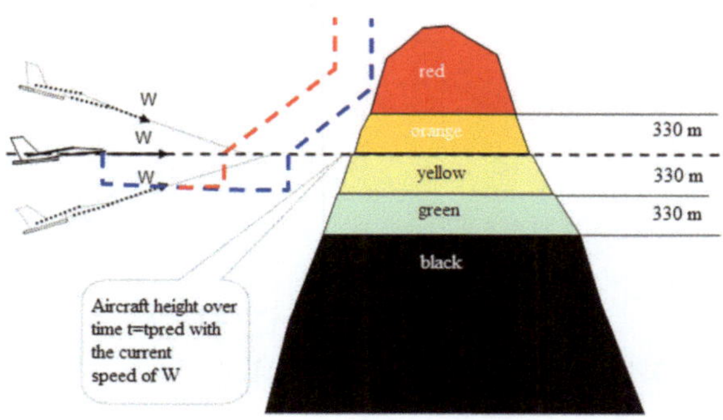

Fig. 4.6 Terrain colors when displayed

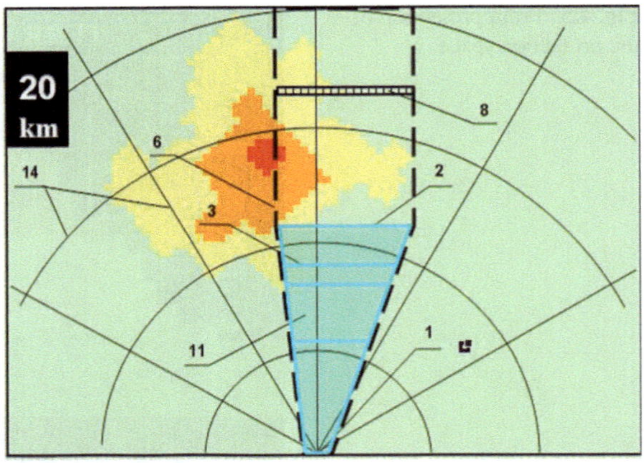

Fig. 4.7 Information scanning region of the terrain relief database for the plan-view projection

In the example in Fig. 4.6, the colors are distributed depending on the elevations of the terrain elements as follows: Terrain elements in the range of elevations from the predicted aircraft height and lower by 330 m are colored yellow; terrain elements below the predicted aircraft height in the range from 330 to 660 m are shown in green color; terrain elements that are 660 m below the predicted aircraft height are colored in black; terrain elements in the range of elevations from the predicted height and 330 m higher are colored in orange; and terrain elements above the predicted aircraft height in the range of elevations of 330 m and more are shown in red.

The cross (see 4 in Fig. 4.4) and front (see 5 in Fig. 4.5) projections of the terrain are synthesized by selecting the maximum heights of the terrain elements presented on the plan-view projection within the corresponding regions of information scanning for terrain elements (see 6 in Fig. 4.7 and 7 in Fig. 4.8).

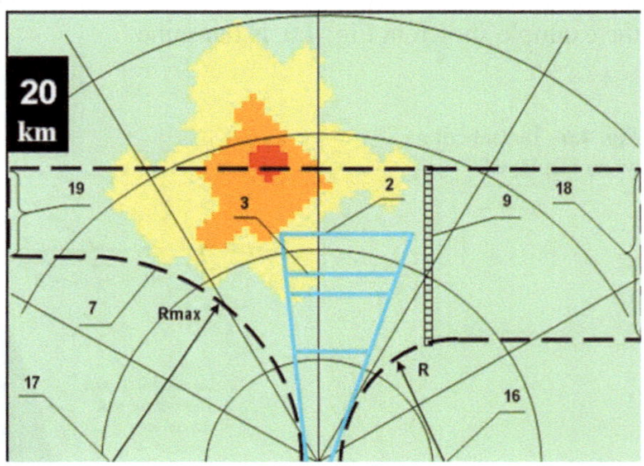

Fig. 4.8 Information scanning region of the terrain relief database for the front projection

For the cross projection, scanning is performed within the lines (see 8 in Fig. 4.7), and for the front projection—within the columns (see 9 in Fig. 4.8) of the corresponding information scanning regions.

The configuration of the information scanning regions for generating the cross 6 and front 7 projections of the terrain shown in Figs. 4.7 and 4.8 is selected based on the calculation of the terrain contours to be displayed on the cross and front projections within the space of the possible aircraft position, taking into account the parameters of its movement, and the scale of the terrain displaying (see 10 in Figs. 4.3, 4.4, 4.5 and 4.6). In addition, the configuration of the information scanning region for generating the terrain cross projection 6 is selected from the condition of coincidence in time of the alert or warning signal with the moment the contour of the synthesized cross projection of the terrain contacts, respectively, with the contour of the PrSp 2 cross projection or the contour of the cross projection of the emergency warning region 3 (see Fig. 4.4).

Such coincidence in time is ensured by the coincidence of the configurations of the PrSp 2 plan-view projection and the scanning region 11 near to the aircraft (see Fig. 4.7) used for synthesizing the cross projection of the terrain. The coincidence in time of the moment of the corresponding alert/warning generation with the moment of contact for the contour of the terrain cross projection and the contour of the PrSp cross projection or the contour of the emergency warning region cross projection makes it possible for the crew not only to respond properly to the signal formed at the moment of contact, but also to use a combined cross projection (see Fig. 4.4) for predicting the probability of a hazardous situation and taking a decision on proactive maneuvering aimed at eliminating such a probability.

The video images of the combined cross, front, and plan-view projections of the terrain and the PrSp agreed in scale allow for a visual alert/warning of the hazard for the crew and provide the crew with a more accurate estimation of the terrain configuration posing hazard for the aircraft by displaying the contact of the PrSp cross projection with the terrain cross projection (see Fig. 4.4) and selecting the terrain elements (see 12 in Fig. 4.4) found inside the PrSp on all formed video images of the combined terrain projections (plan-view, cross and front ones).

According to the combined plan-view, cross, and front projections of the terrain using the appropriate scale grids displayed on the screen (see 13 in Fig. 4.4, 14 in Fig. 4.7 and 15 in Fig. 4.5), the crew estimates the distance between the PrSp projections with the same name and the hazardous terrain zone, along a combined cross projection—that is, decides on the need to correct the path angle Θ (see Fig. 4.2), and on the combined front and plan-view projections—on the need and direction of the lateral maneuver to avoid the hazardous terrain. This estimation is mandatorily performed by the crew when the system generates an alert (or emergency warning) signal, but can (if the workload allows) be performed for preventive purposes in order to take proactive measures in time and avoid the of signal generation.

The information scanning region (see 6 Fig. 4.7) used to form the cross projection is determined as a part of the plan-view projection of the terrain bounded by the PrSp contour (see 11 in Fig. 4.7) with the rectangular zone adjacent to it having a transverse dimension equal to the size of the front PrSp boundary and with a longitudinal dimension equal to the distance from the front boundary to the boundary of the plan-view projection of the terrain (see Fig. 4.7).

To form the front projection (see 5 in Fig. 4.8), the information scanning region is determined with a boundary (see 7 in Fig. 4.8) as two arcs of circles of the right and left turns with radii R and R_{max} with centers 16 (for the right turn arc) and 17 (for the left turn arc) on the line of the lower boundary of the plan-view projection of the terrain (see Fig. 4.8). The arcs mentioned above are closed with a rectangular section of the boundary with a transverse dimension equal to the transverse dimension of the plan-view projection of the terrain with a maximum longitudinal dimension 18 equal to the difference between the distance to the non-hazardous terrain L_{nht} and the arc radius R of the turn being executed, and the minimum longitudinal dimension 19 equal to the difference between the distance to the non-hazardous terrain L_{nht} and the radius R_{max} corresponding to the bank angle γ, the absolute value of which does not exceed 10° (see Fig. 4.8). In this case, centers 16 and 17 are located on the line of the lower boundary of the plan-view projection of the terrain so that the lateral boundaries of the PrSp plan-view projection are tangent to the mentioned arcs of circles the radii of which are defined by the expression:

$$R = W_{gnd}^2 / g\,\mathrm{tg}(\gamma_{calc.}), \tag{4.1}$$

where W_{gnd} is the ground speed (horizontal projection of the general speed vector W), g is the acceleration of gravity, $\gamma_{calc.}$ is the estimated bank angle defined as:

$$\gamma_{calc.} = \begin{cases} \gamma, & if\,\mathrm{ABS}(\gamma) > 10°; \\ 10°, & if\,\mathrm{ABS}(\gamma) < 10° \text{ or no information is available.} \end{cases} \tag{4.2}$$

In this case, non-hazardous terrain means terrain at a safe distance from the aircraft L_{nht} not less than R_{max}, not less than $W_{gnd} \cdot T_{spec}$, not more than $2 \cdot R_{max}$ and not more than the vertical dimension of the plan-view terrain projection.

As shown in Fig. 4.2, the surface bounding, the formed PrSp faces the terrain, is aligned at the initial point (origin) with the aircraft, expands in the direction of the flight, and the distance between its front edge and the aircraft corresponds to the specified flight time T_{spec} with the current ground speed W_{gnd}. The PrSp is expanded symmetrically in the direction of the flight when the aircraft is not maneuvering: to the left and to the right by an angle Δ_{min}. If there is a non-zero turn speed ω_y, the PrSp is expanded asymmetrically in the direction of the flight: by an angle $\Delta_{min} + \Delta$ in the direction of the executed turn and by an angle Δ_{min} in the opposite direction. In this case, the additional angle of expansion of the protective space area Δ is determined depending on the angular speed of the turn performed toward the expansion: $\Delta = f1(\omega_y)$ and is limited in magnitude: $\Delta_{min} + \Delta < \Delta_{max}$.

The algorithm for forming the PrSp boundary surfaces is described in detail in the author's paper [2]. Here it should only be noted that inside the PrSp an alert signal region is calculated that is similar in geometry to the PrSp and geometrically determined by the minimum allowable flight times to the hazardous

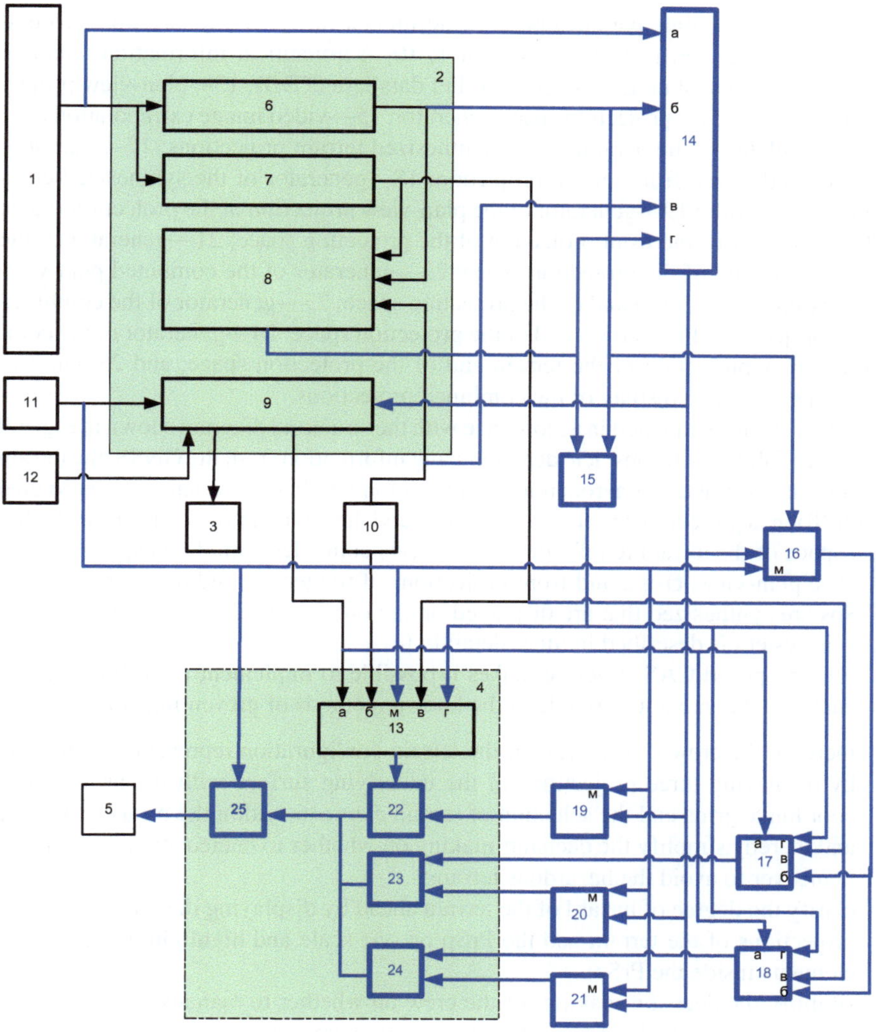

Fig. 4.9 CAS structure for implementation of the method of three-dimensional synthesis of underlying surface sections

terrain (see Fig. 4.2). If the terrain elements defined by the onboard database are located within the alert signal region during the flight, an alert is generated about the hazardous terrain in the direction of flight.

The structure of the collision avoidance system (CAS) implementing the method described above is presented in Fig. 4.9 and is described in detail in [4].

Figure 4.9 has the following designations: 1—navigation system; 2—obstacle detector; 3—alarm device; 4—video generator; 5—display; 6—calculator of the current dynamic state parameters; 7—determinant of the coordinates; 8—calculator of the predicted path; 9—comparator; 10—aeronautical information database (AID); 11—control unit; 12—terrain relief database (TDB); 13—plan-view projection generator; 14—protection space generator; 15—video image extrapolation unit; 16—calculator of the parameters of synthesized terrain projections; 17—generator of the synthesized front terrain projection; 18—generator of the synthesized cross terrain projection; 19—generator of the plan-view projection of the protection space; 20—generator of the front projection of the protection space; 21—generator of the cross projection of the protection space; 22—generator of the combined plan-view projection of the terrain and of the protection space; 23—generator of the combined front projection of the terrain and of the protection space; 24—generator of the combined cross projection of the terrain and of the protection space; and 25—unit of selecting the video display of the combined projections.

The CAS implemented in accordance with the structural diagram shown in Fig. 4.9 works as follows. The obstacle detector using information from the GNSS, the terrain relief database and the aeronautical information database calculates the predicted path. Then, a protection space is constructed, and the comparator determines whether the space of the terrain relief database elements coincides with the PrSp.

The plan-view, cross, and front projections of the terrain and the corresponding PrSps are synthesized that are displayed on the onboard indicator. The functioning of this system is described in more detail in [4].

The proposed CAS structure makes it possible to implement the following new advantages that are not available in the known methods of preventing collisions:

- increase the crew's awareness of the terrain configuration representing a hazard, by displaying three projections of the underlying surface with the use of three coordinate grids and the selection of terrain elements within the protection space and thereby simplify the decision-making on whether to maneuver and to select a maneuver to avoid the hazardous terrain;
- clarify the degree of hazard of the terrain ahead by displaying the three combined projections of the terrain and the PrSp on one scale and highlighting the terrain elements inside the PrSp;
- simplify the decision-making for the crew on whether to maneuver and to select a maneuver to avoid the hazardous terrain by increasing the crew's awareness of the terrain configuration and clarifying the degree of hazard it represents;
- expand the crew's capabilities to prevent hazardous proximity to the terrain by enabling the analysis in three combined projections of the terrain ahead, both in case of hazardous proximity to the terrain (by automatically switching of the display to the terrain display mode), or at any time convenient for the crew (by manual switching of the display to the terrain display mode). Implementation of the above-mentioned advantages makes it possible to increase flight safety by predicting the collision probability and expanding the crew's capabilities to prevent hazardous proximity to the terrain by increasing the amount of information on the position (geometry) of the aircraft and the hazardous terrain (displaying three terrain projections, use of three coordinate grids, displaying the projections

of the protection space, and highlighting terrain elements within the protection space). At the same time, the ergonomics of the information presentation to the crew is enhanced by displaying the combined projections of the terrain and of the protection space, using vertical speed when coloring the terrain relief, and highlighting the terrain elements within the protection space.

Let us evaluate the effect of the developed method on flight safety using the factors defined earlier. As an initial level, we will consider the CAS, in which only the horizontal projection is shown on the display and the PrSp boundaries are not shown. Such systems are currently installed on the majority of civil aircraft throughout the world and a significant portion of aircraft in Russia. An example of these systems is the world's most widely used EGPWS by Honeywell.

If the pilot sees the plan-view projection at the moment of the alert/warning signal triggering, then he/she must assess the situation for making a decision, i.e., to determine the obstacle altitude to decide on the possibility of vertical maneuver, and in case such, an overflight is not possible, to determine the direction of the lateral maneuver or to decide on a reverse turn. In solving this problem, the pilot will have, among other things, to fix his/her eyes on the situation outside the cockpit (external environment).

Numerous experimental studies [16] show that the pilot's attention switches after an alarm/warning triggering within 0.7–3.5 s. The time to request displaying the necessary additional information on the display is 6.5 s on the average. The time of fixing the eyes on the situation outside the cockpit with manual control is 2.7–14.0 s. The time required for the pilot to solve the problem of selecting the necessary maneuver, taking into account a variety of factors and depending on the specific situation, can reach 10–20 s or more, and before the maneuver begins, it may take another 3–8 s. At the same time, in a stressful situation or when performing other tasks, for example, in the conditions of air fight, the probability of making an erroneous decision increases.

Thus, to determine the nature of a safe maneuver when an alert/warning signal triggers in the existing CASs, the pilot will need 22.9 s $(0.7 + 6.5 + 2.7 + 10 + 3)$ to 52 s $(3.5 + 6.5 + 14 + 20 + 8)$.

Obtaining complete information about the nature of the hazard in the direction of flight will allow the pilot to save at least the following time: 6.5 s to switch screens for more information; 2.7 s to focus his/her attention on the external environment; and 5.0 s for decision-making. Thus, the total time spent by the pilot before starting the maneuver is 8.7 s $(0.7 + 5 + 3)$. Then, the ergonomics factor F_{Erg} can be calculated as the ratio of the time for making a decision on a specific maneuver with the use of the proposed method to the time that would have been spent without using it. The above data show that these times can be estimated as 8.7 and 22.9 s, respectively, and $F_{\text{Erg}} = 22.9/8.7 \approx 2.6$.

To estimate the reduction of the probability of an accident (F_{Acc}), we estimate the increment in height that an aircraft can gain by saving time for the pilot to make a decision on a maneuver. For example, consider civil aircraft, for which a constraint on comfortable overload of $0.25 * g \approx 2.5$ m/s^2 is provided, and the vertical component of the speed does not exceed 20% of the longitudinal speed to prevent the aircraft from entering the airflow breakdown mode.

Without violating the approach, we will assume that at the moment of the alert/warning signal triggering the aircraft was in the level flight mode. Then, it is not difficult to calculate the height that the aircraft can gain over the obstacle signaled when the latter appears in the protection space. For civil aircraft, an alarm/warning will appear 45 min from the flight start to the obstacle (see Fig. 4.2). Using the method discussed above, climbing will begin in 8.7 s, and if the longitudinal speed is assumed to be 100 m/s and it will not increase during the maneuver, it is easy to calculate that by the time of approaching to the obstacle, the aircraft will gain additional 726 m ($100 \cdot 0.2 \cdot (45 - 8.7)$) of the height. A similar calculation, taking into account the delay of 22.9 s, shows that the aircraft will only gain additional 442 m ($100 \cdot 0.2 \cdot (45 - 22.9)$) and then $F_{\text{Acc}} = 726/442 \approx 1.6$.

Thus, the above method makes it possible for the crew to properly estimate the degree of hazard and to make the right decision about the need and nature of maneuvering. The expert estimate shows that the proposed method allows saving up to ≈ 14 s ($22.9 - 8.7$) for the pilot to make a decision on how to maneuver; therefore, the aircraft has the chance to gain 1.6 times the height when flying over the obstacle.

Next, we will consider a method that allows for further improvement of flight safety, especially for military aircraft performing low-altitude flights under the conditions of hazardous terrain, including combat missions.

4.2 Method for Increasing the Flight Effectiveness and Safety by Assessing the Possibility of Vertical Maneuvering and Determining the Direction of the Turn

As was shown in the previous section, the method for forming three projections of the underlying surface provides the aircraft pilot with the most complete picture of the nature of the terrain in the direction of flight. At the same time, at every particular moment in time, especially when performing flights over the hazardous terrain of the underlying surface, the task of selecting the safest maneuver requires additional time from the pilot to make a decision, which can lead to an accident.

A drawback of the known methods and systems [2, 3, 13, 17, 18] that provide to solve similar tasks is that in the event of a hazardous situation, a collision alert/warning is only made by analyzing the possibility of vertical maneuvering. The negative result of such an analysis (the impossibility of preventing a collision by vertical maneuvering) only indicates that a lateral maneuver is necessary. However, the direction and trajectory of the lateral maneuver are not determined and are left to the pilot's discretion.

The task of this method is not only to assess the possibility of vertical maneuver, but also, with a negative result of such an assessment, to determine the direction of

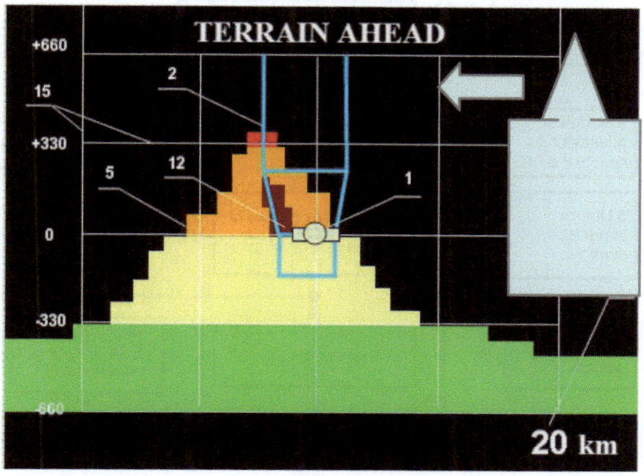

Fig. 4.10 Turn direction assessment

the turn by analyzing both the left and the right turns, and selecting the one most preferable in this situation.

The idea of the proposed method consists in an integrated approach to selecting the safest path, taking into account the constraints imposed by the nature of the terrain, the aircraft performance and other available information. For illustration, Fig. 4.10 presents the front projection of the terrain where the no-flight zone is shown. If it is impossible to perform a vertical maneuver, the pilot may decide to fly around the obstacle from the right. The CAS, whose database contains information about the no-flight zone or information from the weather radar for civil aircraft, can offer a fly around from the left, as it is safer in this particular situation. By this the CAS shortens the time for the pilot to decide on the nature of the maneuver required and thus enhances flight safety in general.

The method is explained in Fig. 4.11, which presents all the functional elements necessary for its implementation, the main of which are: the GNSS as a source of navigation information, a terrain relief database, a vertical flyby unit, a turn direction block, an onboard display, and an aircraft performance module. In the method under consideration, the navigation system calculates the parameters of the current aircraft dynamic state, estimates its coordinates with extrapolation over a predetermined time interval, calculates the predicted path, compares it with the terrain, and warns of hazard by generating alarms and video images of the hazardous terrain. The interaction of the functional elements necessary for the implementation of the method under consideration is described in more detail in [3, 5].

Thus, in analogous methods, when these are implemented in specific devices, only a signal about the results of the vertical maneuver analysis is provided and, if the result is negative, another signal is provided about the need for lateral maneuver. However, the direction and necessary parameters of the lateral maneuver are not specified.

In the proposed method (see Fig. 4.11), this function is performed by the turn direction block 16, which includes the flyby condition analyzer 17 that receives a

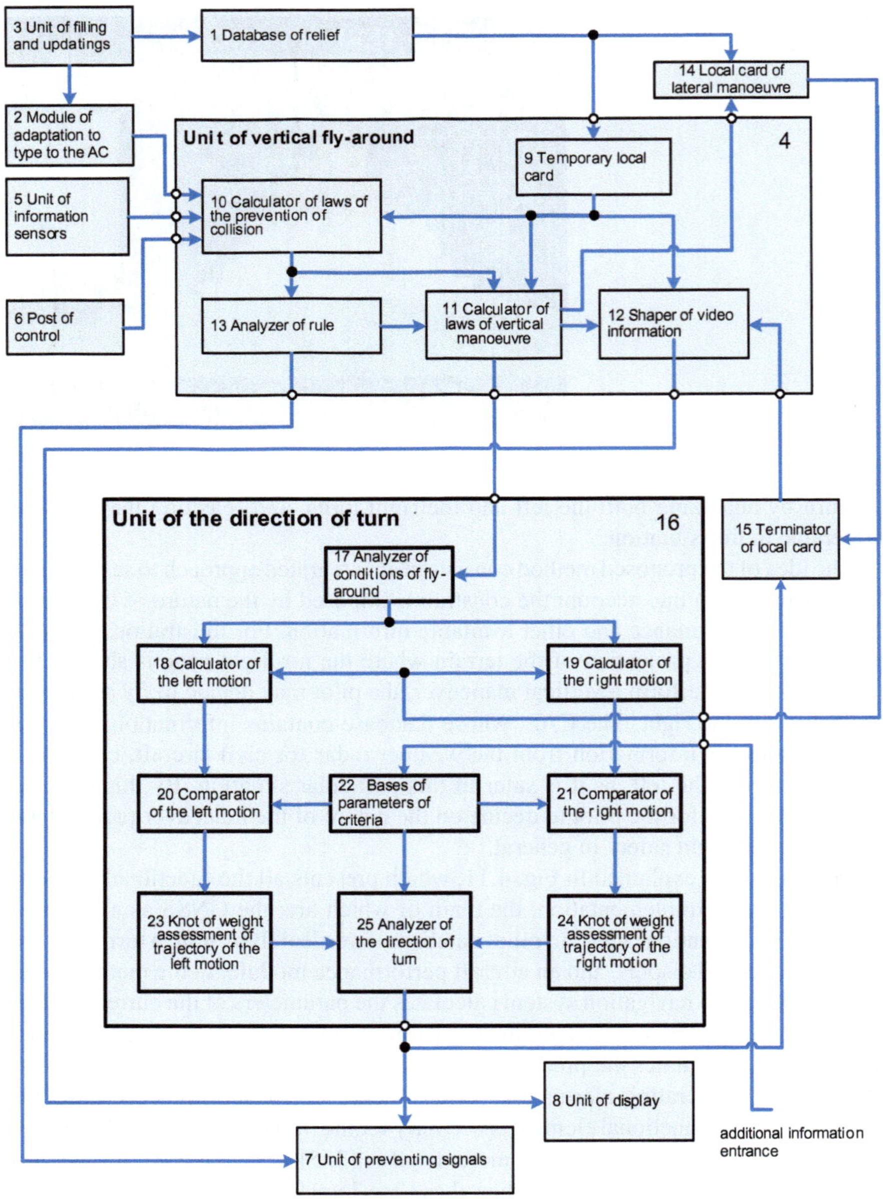

Fig. 4.11 Method for determining the turn direction

vertical maneuver hazard indicator and the current parameters of the aircraft state vector at its input from the vertical flyby block 4 (from the output of the vertical maneuver law calculator 11). In the analyzer 17, based on the aircraft state parameters (coordinates, speed, acceleration, angles, etc.) normalized data are generated for transmission to the calculators 18 and 19.

In the above calculators, all possible trajectories of lateral maneuvers are generated taking into account the aircraft dynamic and ergonomic, and these data are matched with the terrain data in accordance with the local map 14 coming to the inputs of the calculators 18 and 19.

The set of generated data packets comes from the calculators 18 and 19 to the inputs of the comparators 20 and 21. The criteria parameter database 22 contains continuously updated data on the minimum allowable altitudes of the aircraft, ergonomic (maximum allowable overloads) and dynamic characteristics of the aircraft, data from air traffic control (ATC) systems (presence of aircraft from opposed or in-trail directions, width of the flight route, restricted flight areas, etc.), from an onboard weather radar (presence of weather hazards the aircraft flight area) coming through an additional data input. The data from the database 22 are fed to the inputs of the comparators 20 and 21, in which the formed trajectories of the lateral maneuvers are compared with the data from the local lateral maneuver map coming through the calculators 18 and 19; at the same time, the optimality indicator is formed for each trajectory according to the criteria of minimum deviation from the flight plan, the maximum vertical clearance from the aircraft to the earth's surface and the horizontal distance between the aircraft and the obstacle when flying along the intended trajectory.

From the comparators 20 and 21, the calculated possible trajectories of the lateral maneuvers with optimality indicators obtained on the basis of the comparative analysis performed in the comparators, are fed to the corresponding weight estimation units 23 and 24.

In the weight estimation units, the optimality parameters of all formed trajectories are compared taking into account the data contained in the criteria parameter database and only one trajectory from this number of analyzed trajectories is selected that meets the given criteria to the maximum extent possible.

The trajectories of the left and right traffic selected at the units 23 and 24 are transferred to the turn direction analyzer 25, where the most preferred direction of the turn and the corresponding trajectory are finally selected using the data from the criteria parameter database 22. The result of the selection comes from the analyzer 25 to the alert/warning signal unit 7 that outputs a signal to the pilot to perform the appropriate actions to establish on the selected path. In addition, the signal from the analyzer 25 enters the local map limiter 15, which selects the map section corresponding to the selected path for further display. In this case, the formation of video information for the display unit 8 is performed by the generator 12, where the information from the temporary local card 9 is replaced over the signal from the additional output of the calculator 11 with information from the local card limiter 15 and is then transmitted from the output of the generator 12 to the input of the display unit 8. The signals that earlier arrived to the display unit 8 from the output of the calculator 11 are blocked.

Thus, due to the block of algorithms for determining the direction of the turn in the proposed method, if vertical maneuver is impossible, not only a signal (indicator) of the need for lateral maneuver is formed, as in other known CASs, but also the direction and trajectory of the turn are determined by analyzing both the left and the right turn, and the most preferable of them is selected in this situation.

Let us evaluate the effect of the proposed method on flight safety using the factors defined in 1.6. As an initial level, we will consider the CAS, in which the method of determining the direction of the turn is not used when an alert/warning signal appears.

If the pilot sees the plan-view projection of the underlying surface at the moment of the alert/warning signal triggering, then he/she must assess the situation for making a decision, i.e., to determine the obstacle altitude to decide on the possibility of vertical maneuver, and in case such, an overflight is not possible, to determine the direction of the lateral maneuver or to decide on a reverse turn. In solving this problem, the pilot will have, among other things, to fix his/her eyes on the situation outside the cockpit (external environment).

As noted earlier, experimental studies [16] show that the pilot's attention switches after an alarm/warning triggering within 0.7–3.5 s. The time to request displaying the necessary additional information on the display is 6.5 s on the average. The time of fixing the eyes on the situation outside the cockpit with manual control is 2.7–14.0 s. The time required for the pilot to solve the problem of selecting the necessary maneuver, taking into account a variety of factors and depending on the specific situation, can reach 20–30 s. At the same time, in a stressful situation or when performing other tasks, for example, in the conditions of air fight, the probability of making an erroneous decision increases.

The block of algorithms for determining the direction of the turn, which takes into account a lot of information about the nature of the underlying surface, weather conditions, ATC data, civil aircraft performance characteristics, allows expert estimating the amount of time saved for deciding on the maneuver of at least 20 s. At the same time, the ergonomics factor F_{Erg} can be calculated as the ratio of the time for making a decision on a flyby maneuver with the use of the proposed method to the time that would have been spent without using it, considering the fact that in order to make such a decision, the pilot will need at least the following time:

3 s for attention switching;
$n * 6.5$ s to request the help page with additional information on the display screen, where n is the number of requests of various pages;
10 s for visual analysis of the external environment;
20 s for decision-making.

Thus, assuming $n = 2$ (i.e., the pilot will analyze two information pages on the display when using the conventional CAS), we obtain an estimate of the total time before starting the maneuver equal to 46 s. The use of the CAS, in which the proposed method for determining the direction of the turn is implemented, will require 3 s without visual analysis of the external environment or 13 s with additional visual analysis. In this case, the minimum value of the ergonomics factor can be estimated as $\mathbf{F_{Erg}} = 46/13 \approx 3.5$.

To estimate the reduction in the probability of an accident F_{Acc}, we estimate the additional distance to the obstacle that the pilot of the aircraft will gain performing the turn maneuver after receiving the CAS instructions, or by making a decision independently, but 20 s later. Let us assume for definiteness that the ground speed of the aircraft is $V = 200$ m/s.

Then, when using the CAS, in which the considered method of determining the direction of the turn is implemented, we obtain the distance flown in the direction of the obstacle until the decision on the maneuver is made $200 * 13 = 2{,}600$ m, and if the pilot makes his/her own decision, we obtain a distance of $200 * 46 = 9{,}200$ m. From this we obtain the estimate: $F_{Acc} = 9200/2600 \approx 3.5$.

Thus, the method of determining the direction of the turn considered above makes it possible for the crew to significantly shorten the time for making a decision on the appropriate maneuver. The expert estimate shows that the proposed method allows saving up to 20 s for the pilot to make a decision to perform a maneuver, and thus the aircraft has a margin of more than three times the distance to the obstacle, which increases flight safety. It should also be noted that these estimates do not take into account possible erroneous decisions of the pilot, especially in a stressful situation.

Further, we will consider one more method which allows further improvement of flight safety, especially for low-altitude flights under the conditions of the hazardous terrain.

4.3 Method for Increasing the Flight Effectiveness and Safety by Identifying Hazardous Terrain, Taking into Account the Possibility of a Reverse Turn, and the System Structure for Its Implementation

In the above methods for improving flight safety with the use of the CAS, various options were proposed based on improving the ergonomic characteristics of displaying the current situation on the onboard displays or on advising the pilot about the necessary maneuvers. Further development of the considered direction of the CAS improvement, as noted earlier (see Sect. 1.5, Fig. 1.1), is preventing the aircraft from encountering potentially hazardous situations.

The idea of the proposed method consists in a preliminary calculation of the potential hazard of the flight along the intended path and is illustrated in Fig. 4.12, which shows the plan-view projection of the underlying surface and the predicted flight path of the aircraft. If a flight along this path does not allow a reverse turn maneuver to be performed at any time, then such a path is potentially hazardous. An example is a flight in mountain conditions, when there is no possibility of climbing.

The drawback of all known methods [4, 5, 13–15, 19, 20], including those described above, is the fact that there are situations in which it is impossible to avoid

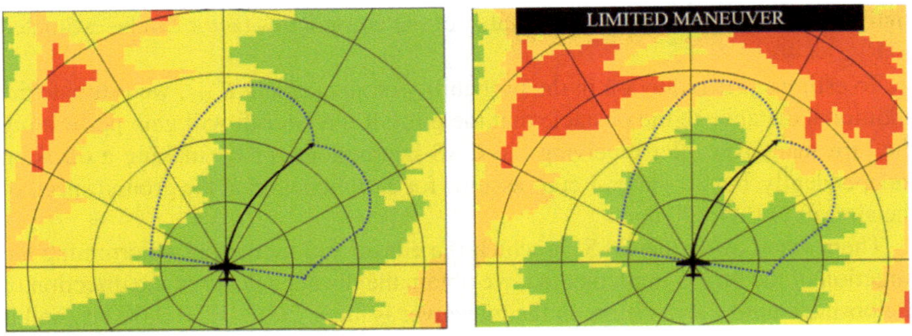

Fig. 4.12 Preventing the aircraft from encountering potentially hazardous situations

the collision with the terrain. This drawback is always present due to the fact that the proximity of the hazardous terrain is determined strictly along the aircraft heading in the range of the aircraft positions (scatter). However, there may be situations where, if some hazardous terrain is detected along the heading and it is impossible to avoid the hazardous terrain by climbing, the aircraft will have to perform a reverse turn.

In this case, it is necessary to provide a safe space for turning in one or another direction. As is known [21–23], the permissible turn radius is a variable value that depends on the aircraft dynamic parameters at the given moment.

If the hazardous terrain is, for example, a ravine, an alarm may be output when estimating a lateral turn in any direction. Consequently, in the case of a negative estimation of the vertical maneuver possibility, a lateral turn will also be impossible and an accident will be inevitable.

Thus, the drawback mentioned above is inherent in all known methods of preventing collisions with the terrain.

The objective of the proposed method is to improve flight safety by identifying hazardous terrain taking into account the possibility of a reverse turn (to the right or to the left of the predicted path).

In the proposed method, the boundaries of the safe corridor are determined by scanning the space with the predicted path line to the left and right of the path, and the width of the safe corridor is defined by the formula:

$$L = 2 \times \left[R_{\text{pr.rt}} + R_{\text{pr.lt}} + e_1 + e_2\right], \tag{4.3}$$

where L is the width of the safe corridor; $R_{\text{pr.rt}}$ is the predicted minimum permissible radius of the right turn; $R_{\text{pr.lt}}$ is the predicted minimum permissible radius of the left turn; e_1 is the maximum error in positioning of the aircraft; and e_2 is the minimum safe lateral distance of the aircraft from the terrain.

The idea of the proposed method is explained with the help of Fig. 4.13. In the proposed method, the navigation system determines the aircraft position, parameters of the current dynamic state of the aircraft (ground speed W_{gnd}, vertical speed W_y, track angle TA, turn rate ω_y, etc.); these are used to extrapolate the aircraft position over the predetermined time interval (T_{fl}) and to calculate the predicted path. Then the minimum permissible turn radii (both to the right and to the left of the predicted path) are calculated based on the dynamic state parameters and the predicted (at the time the hazardous terrain is detected) turn radii based on the calculated predicted dynamic parameters of the aircraft; then a boundary of the safety tunnel (ST) is determined, within which the predicted path must be located.

The ST boundaries are determined on the basis of the reverse turn possibility in accordance with the predicted minimum permissible turn radii. The ST comparison with the terrain (within the formed sample from the terrain relief database) is performed by scanning the space within the ST. In the case a part of the terrain coincides with the ST, an alert/warning on hazard is generated.

The ST is shown in Fig. 4.14 and represents a space directly linked with the current aircraft position (see 11 in Fig. 4.14) and determined (by the form and attitude) by the predicted path 12, the minimum permissible radii of turn circles 13 (to the right) and 14 (to the left) and predicted (for a predetermined time interval) minimum permissible radii of the turn circles 15 (to the right) and 16 (to the left). In addition, the ST is formed based on the maximum error in determining the aircraft position e_1 and the minimum safe lateral distance between the aircraft and the terrain e_2 and is also determined by the movement parameters (W_{gnd}, W_y, TA, ω_y), the minimum allowable height H_{ma} and the additional height margin ΔH determined from the condition of minimizing the probability of false alarms of the notification system. The H_{ma} value, as is known [24], depends on the flight phase performed (cruise flight, flight in the terminal area, approach) and flight mode (level flight, descent, climb). The sum ($H_{ma} + \Delta H$) is the distance 17 by which the lower part of the ST boundary 18 is offset from the predicted path and the sum of values ($e1 + e2$) 19 is the minimum distance from the ST boundary to the turn circles both with minimum allowable radii and with the predicted minimum allowable radii.

The ST boundary configuration 18 is recalculated with a rate required to update the information, and thus, the ST with a width of 20 is adapted to the current dynamic state of the aircraft and to the flight mode. This technique greatly facilitates the crew's task to make a decision in the current situation, as it allows evaluation of the degree of hazard when approaching the terrain before a dangerous situation emerges, and does not require the preventive view of the terrain display, i.e., it reduces the psychological stress of the crew.

Thus, it is possible to avoid a hazardous situation onboard the aircraft (alarm generation) by timely maneuvering performed on the basis of the generated signal notifying of a potentially hazardous terrain.

The cross projection of the terrain (21 in Figs. 4.14b and 4.16) is synthesized by selecting the maximum elevations of the terrain elements presented on the plan-view projection, within the boundaries of the information scanning region for the terrain elements (22 in Figs. 4.14a and 4.15). To generate a cross projection, scanning is performed within the lines of the information scanning region (23 in Fig. 4.15).

The configuration of the information scanning area with boundaries 22 for generating the cross projection of the terrain 21 shown in Figs. 4.14b and 4.16, respectively, is selected depending on the displaying the terrain outline located within the ST space

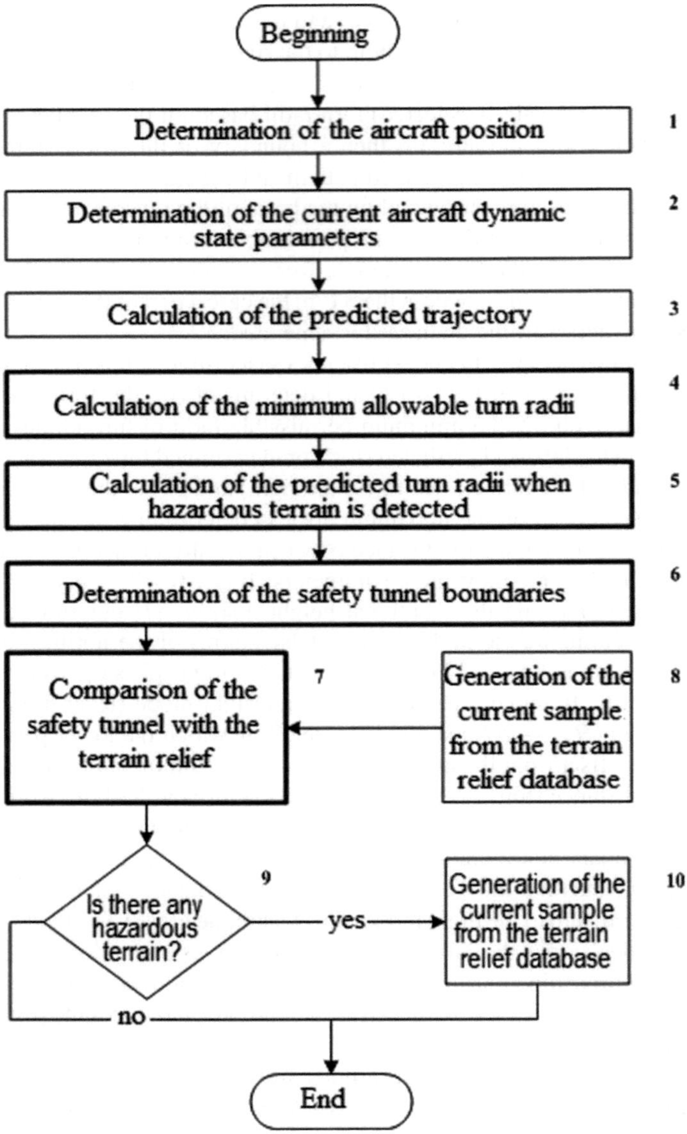

Fig. 4.13 Sequence of operations in the method of preventing aircraft collisions

(a)

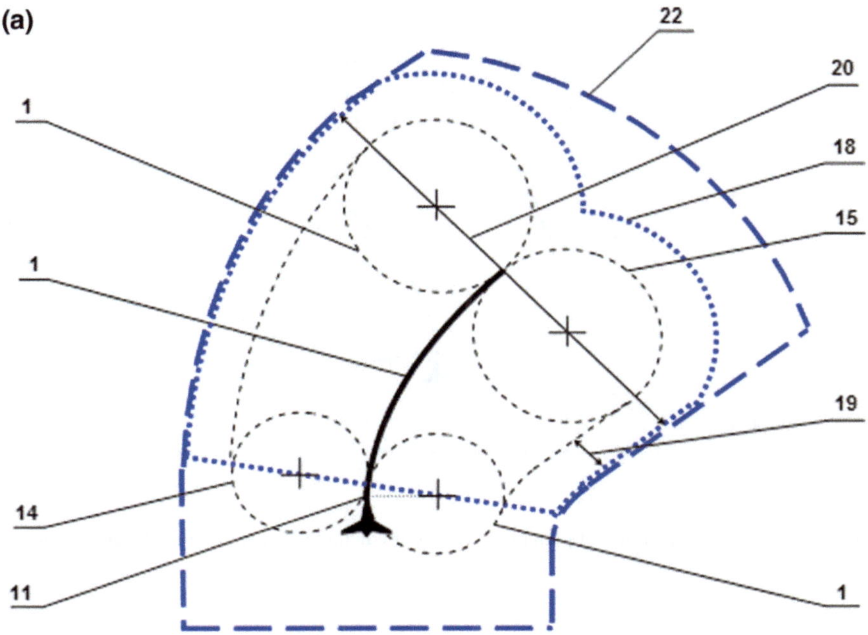

(b)

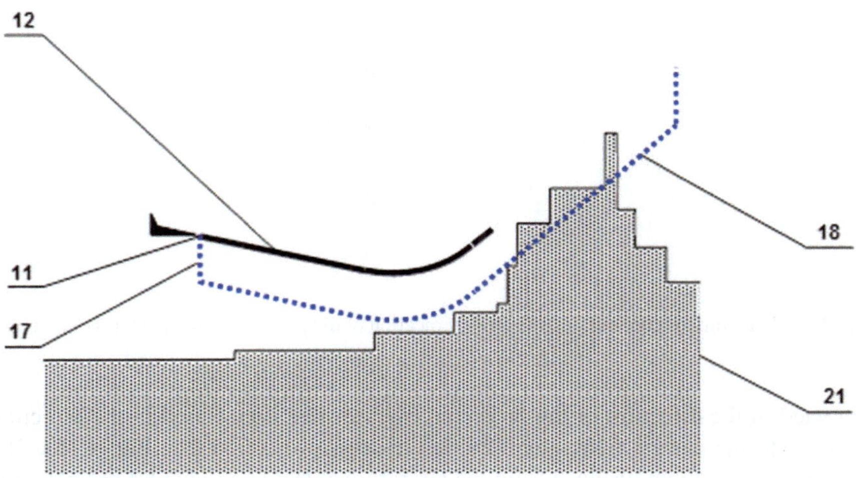

Fig. 4.14 Plan (**a**) and profile (**b**) of projection of predictables trajectory and the safety tunnel

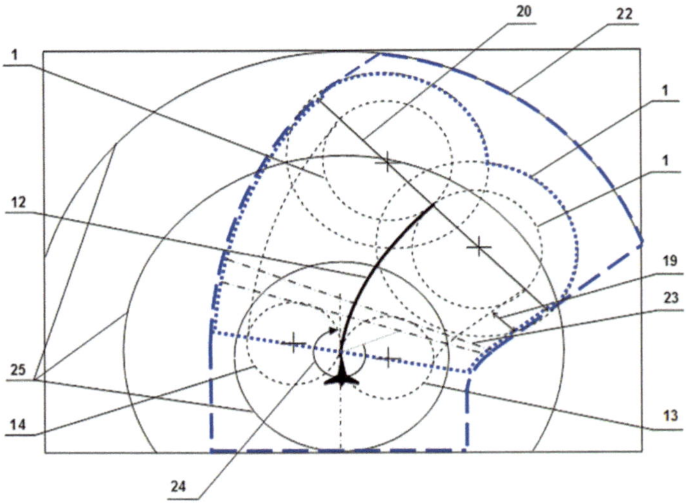

Fig. 4.15 Scheme of creation of plan projection of predictables trajectory and the safety tunnel

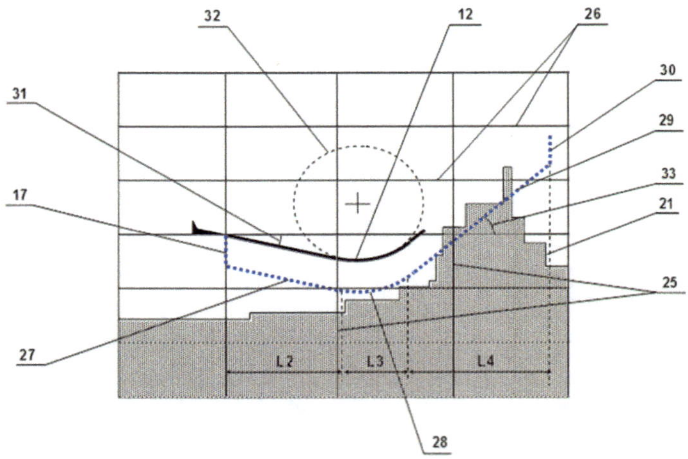

Fig. 4.16 Diagram for constructing the cross projection of the predicted path and the safety tunnel

extended in the direction of the predicted flight of the aircraft and in the direction opposite to the current track angle 24 on the cross projection (see Fig. 4.15). The terrain at a maximum distance from the aircraft displayed on the cross projection is determined by the vertical scale of the plan-view projection range, as can be seen from the comparison of the locations of the range scale grids 25 on the plan-view and cross projections (see Figs. 4.15 and 4.16). The elevation of the hazardous terrain over the aircraft can be estimated with the scale grid 26 (see Fig. 4.16).

As shown in Figs. 4.15 and 4.16, the surface bounding, the formed ST faces the terrain located in the direction of the predicted flight, is aligned at the initial point (origin) with the aircraft, expands (or narrows) in the direction of the flight, and the distance between its front edge and the aircraft corresponds to the position of the turn circle at the maximum distance from the aircraft (in Figs. 4.14b and 4.16, the left circle is at the maximum distance from the aircraft) with the predicted ground speed $W_{predgnd}$. The ST extension (narrowing) in the direction of the flight when the aircraft is not maneuvering along the heading is symmetrical in accordance with the predicted increase (decrease) in the current ground speed W_{gnd}. If there is a non-zero turn speed ω_y, the ST is predicted to curve in the direction of the turn to be performed.

The ST boundary surface (18 in Figs. 4.14 and 4.15) is formed by five surfaces (faces) (17, 27–30 in Fig. 4.16). The algorithm for the formation of these surfaces is described in detail in [1–3]. Terrain with hazardous elevations is understood as the terrain inavoidable at the minimum allowable height above it, when performing a turn with climbing while observing the requirements for permissible vertical and lateral overloads. The first face 17 is vertical, its width is determined by the accuracy of the aircraft positioning ($2e_1$), the safe distance from the terrain ($2e_2$), the current value of the minimum permissible radii of the right ($2R_{pr.rt}$) and the left ($2R_{pr.lt}$) turn, and the height is equal to $H_{ma} + \Delta H$, that is, as a function of H_{ma}, it corresponds to that flight phase and mode to which the ST is adapted. The lengths of the horizontal projections of the second 27 and third 28 surfaces ($L2$ and $L3$, respectively) are determined by the time needed to perform the vertical maneuver of the aircraft when avoiding the hazardous terrain (taking into account the permissible vertical overload) determined by the predicted value of the ground speed $W_{predgnd}$. The slope of the second face 27 relative to the horizon is equal to the path angle Θ 31, the tangent of which is equal to the ratio of the current values of the vertical (W_y) and ground (W_{gnd}) speeds. The third surface 28 is conical or (in the absence of a predicted lateral maneuvering) cylindrical and the length of its horizontal projection $L3$ is determined by the condition of permissible vertical overload for the predicted flight of the aircraft along the arc of the circle 32 lying in the vertical plane. The fourth surface 29 with the horizontal projection length $L4$ rises above the allowable path climb angle equal to Θ_{all} 33, to achieve a ST-controlled distance from the aircraft determined by the predicted flight time T_{fl} with the speed varying from the current value of W_{gnd} to the predicted $W_{predgnd}$ and also determined by the position of the most distant turn circle with the predicted minimum permissible radii ($R_{pr.rt}$ and $R_{pr.lt}$), values of e_1 and e_2 and bounded by the vertical fifth surface 30. If the calculations show that elements of the terrain determined by the formed ST sample, are located within the ST (see Fig. 4.13), an alarm/warning signal is generated.

The minimum permissible turn radii required for the ST are calculated in accordance with the equations of movement, the complexity of which may differ significantly, depending on the flight phase. In the general case (without taking into account the orbital motion of the Earth around the Sun, the diurnal motion of the Earth and the fuel burnout), the minimum permissible radii are determined numerically by solving the system of equations for the aircraft movement given below [21].

The equation of forces in the trajectory coordinate system for aircraft has the following form:

$$\frac{\mathrm{d}}{\mathrm{d}t}(m\overline{\mathbf{V}}_{\kappa}) = \begin{pmatrix} \frac{\mathrm{d}}{\mathrm{d}t}(mV_{\kappa x}) \\ \dot{\Theta} m V_{\kappa x} \\ -\dot{\Psi}\cos\Theta m V_{\kappa x} \end{pmatrix}$$
$$= A_{\mathrm{T}\leftarrow\mathrm{E}} \cdot A_{\mathrm{E}\leftarrow\mathrm{FIX}} \cdot (\overline{\mathbf{P}})_{\mathrm{FIX}} + A_{\mathrm{T}\leftarrow\mathrm{E}} \cdot A_{\mathrm{E}\leftarrow WC} \cdot (\overline{\mathbf{R}}_A)_{\mathrm{WC}}$$
$$+ A_{\mathrm{T}\leftarrow\mathrm{E}} \cdot (\overline{\mathbf{G}})_{\mathrm{E}} + A_{\mathrm{T}\leftarrow\mathrm{E}} \cdot A_{\mathrm{E}\leftarrow\mathrm{FIX}} \cdot (\overline{\mathbf{F}}_{\mathrm{gear}})_{\mathrm{FIX}}, \tag{4.4}$$

where m is the aircraft mass; t is the time; $\overline{\mathbf{V}}_{\mathrm{coor}}$ is the velocity vector of the aircraft in a coordinate system related to the Earth; V_{coorx} is the projection of the earth's velocity on the X-axis of the trajectory coordinate system; Θ is the angle of the trajectory inclination; Ψ is the track angle; symbol $(\overline{\boldsymbol{T}})_i$ indicates the ith vector if the aircraft propulsion $(\overline{\boldsymbol{P}})$, aerodynamic force $(\overline{\boldsymbol{R}}_A)$, gravity $(\overline{\boldsymbol{G}})$, the resultant force of the gear reaction $(\overline{\mathbf{F}}_{\mathrm{gear}})$) in the ith coordinate system; indices "I" denote: FIX—fixed CS, T—trajectory CS, E—Earth CC, WC—wind CS; the symbol $A_{\mathrm{E}\leftarrow\mathrm{FIX}}$ denotes the transfer matrix of the coefficients from one CS to another (in this case, from the fixed coordinate system to the earth one).

The equation of the angular momentum relative to the aircraft center of mass is as follows:

$$\frac{\mathrm{d}}{\mathrm{d}t}(I_{\mathrm{A}} \cdot \overline{\omega}) = (\overline{M}_{\mathrm{P}})_{\mathrm{FIX}} + A_{\mathrm{FIX}\leftarrow\mathrm{WC}} \cdot (\overline{M})_{\mathrm{WC}} + A_{\mathrm{FIX}\leftarrow\mathrm{WC}} \cdot (\overline{r}_{\mathrm{gear}} \times \overline{G})_E$$
$$+ (\overline{r}_{\mathrm{gear}} \times \overline{F}_{\mathrm{gear}})_{\mathrm{FIX}}, \tag{4.5}$$

where $I_{\mathrm{A}} = \begin{pmatrix} I_x & -I_{xy} & -I_{xz} \\ -I_{xy} & I_y & -I_{yz} \\ -I_{xz} & -I_{yz} & I_z \end{pmatrix}$—is the inertia tensor (a symmetric matrix of the aircraft moments of inertia relative to the axes of the fixed coordinate system in which for aircraft having a vertical plane of symmetry, $(I_{xz} = I_{yz} = 0)$; $\overline{\boldsymbol{\omega}} = \{\omega_x; \omega_y; \omega_x\}$ is the vector of the aircraft angular rotation rate, and the matrix $A_{\mathrm{FIX}\leftarrow\mathrm{E}}$ is obtained from $A_{\mathrm{E}\leftarrow\mathrm{FIX}}$ by the conjugation operation, $\overline{\boldsymbol{M}}_{\mathrm{P}}$ is the thrusting (propulsion) torque of the aircraft, $\overline{\boldsymbol{M}}$ is the aerodynamic moment, $\overline{\mathbf{r}}_{\mathrm{gear}}$ is the arm of the gear reaction net force, and $\overline{\boldsymbol{r}}_{\mathrm{G}}$ is the arm of gravity.

Equations of kinematic connections of linear velocities:

$$\begin{pmatrix} \dot{x}_{\mathrm{g}} \\ \dot{y}_{\mathrm{g}} \\ \dot{z}_{\mathrm{g}} \end{pmatrix} = \begin{pmatrix} \dot{L} \\ \dot{H} \\ \dot{z}_{\mathrm{g}} \end{pmatrix} = A_{E\leftarrow T} \cdot \overline{V}_{\mathrm{coor}} = A_{E\leftarrow T} \cdot \begin{pmatrix} V_{\mathrm{coor}} \\ 0 \\ 0 \end{pmatrix} = \begin{pmatrix} V_{\mathrm{coor}}\cos\Theta\cos\Psi \\ V_{\mathrm{coor}}\sin\Theta \\ -V_{\mathrm{coor}}\cos\Theta\sin\Psi \end{pmatrix}, \tag{4.6}$$

where $x_g,\ y_g,\ z_g$ are the aircraft coordinates in the earth CS; L and H are range and altitude of the flight.

Equations of kinematic relations of angular velocities:

$$\begin{pmatrix} \omega_{x1} \\ \omega_{y1} \\ \omega_{z1} \end{pmatrix} = \begin{pmatrix} \dot{\gamma} + \dot{\psi} \sin \vartheta \\ \dot{\psi} \cos \vartheta \cos \gamma + \dot{\vartheta} \sin \gamma \\ \dot{\vartheta} \cos \gamma - \dot{\psi} \cos \vartheta \sin \gamma \end{pmatrix} \quad \text{or} \quad \begin{pmatrix} \dot{\vartheta} \\ \dot{\gamma} \\ \dot{\psi} \end{pmatrix} = \begin{pmatrix} \omega_y \sin \gamma + \omega_z \cos \gamma \\ \omega_x - \mathrm{tg}\vartheta (\omega_y \cos \gamma - \omega_z \sin \gamma) \\ \sec\vartheta (\omega_y \cos \gamma - \omega_z \sin \gamma) \end{pmatrix}, \tag{4.7}$$

where $\omega_{x1}, \omega_{y1}, \omega_{z1}$, are projections of the angular velocity of the aircraft rotation on the axis of the fixed coordinate system; γ is the bank angle; ψ is the yaw angle; and ϑ is the pitch angle.

General assumptions (symmetricity of engine thrust (power), atmosphere at rest) and various assumptions that are possible in trajectory problems on most segments of the performed flight (coordinated flight in a vertical plane, level straight line flight, level steady-state flight, straight climb or descent, constant speed climb or descent, pre-landing descent along the glide path, wings-level level flight with sideslipping, banked level flight without sideslipping) make it possible to reduce with the system (4.4)–(4.7)) in each of these cases to a substantially simpler form [22].

The use of simplified equations for determining the minimum permissible turn radii and calculating the predicted path makes it possible, if necessary, to reduce computational costs and increase the speed of the computations performed.

Thus, the equations for a steady-state curvilinear flight in the fixed coordinate system, if the angular velocities are considered to be small and their products can be neglected, can be written (neglecting the curvature of the earth's surface) as follows:

$$m_x^{\beta}\beta + m_x^{\delta_{н}}\delta_{н} + m_x^{\delta_{э}}\delta_{э} + m_x^{\overline{\omega}_x}\overline{\omega}_{x1} + m_x^{\overline{\omega}_y}\overline{\omega}_{y1} = 0, \tag{4.8}$$

$$m_y^{\beta}\beta + m_y^{\delta_{\mathrm{H}}}\delta_{\mathrm{H}} + m_y^{\overline{\omega}_x}\overline{\omega}_{x1} + m_y^{\overline{\omega}_y}\overline{\omega}_{y1} = 0, \tag{4.9}$$

$$c_z^{\beta}\beta + c_z^{\delta_{\mathrm{H}}}\delta_{\mathrm{H}} + c_y \sin\gamma \cos\gamma \cos\vartheta = -\frac{2m}{\rho S V}(\alpha\omega_{x1} + \omega_{y1}) = -\frac{4m}{\rho S l}(\alpha\overline{\omega}_{x1} + \overline{\omega}_{y1}), \tag{4.10}$$

$$m_z^{\alpha}\alpha + m_z^{\delta_{\mathrm{B}}}\delta_{\mathrm{B}} + m_z^{\overline{\omega}_z}\overline{\omega}_{z1} = 0, \tag{4.11}$$

where β is the sideslip angle; δ_{Rudder} is the angle of the rudder deflection; δ_{Ail} is the aileron deflection angle; $\overline{\omega}_{x1}, \overline{\omega}_{y1}, \overline{\omega}_{z1}$ are unitless projections of the angular velocity of the aircraft on the axis of the fixed CS; m_x^{β} is the partial derivative of the bank aerodynamic moment coefficient determined by the aerodynamic parameter (in this case, sidesliding); m_y^{β} is the partial derivative of the yaw aerodynamic moment coefficient determined by the aerodynamic parameter (in this case, sidesliding); c_z^{β} is the partial derivative of the cross-stream force coefficient determined by the aerodynamic parameter (in this case, sideslipping); c_y is the aerodynamic normal force coefficient; ρ is the air density; S is the area of the equivalent wing; V is the upstream velocity; α is the angle of attack; and l is the wing span.

For small pitch angles, expressions (4.7) become:

$$\left.\begin{aligned}\omega_{x1} &= \dot{\psi}\sin\vartheta \approx \dot{\psi}\vartheta,\\ \omega_{y1} &= \dot{\psi}\cos\vartheta\cos\gamma \approx \dot{\psi}\cos\gamma,\\ \omega_{z1} &= -\dot{\psi}\cos\vartheta\sin\gamma \approx -\dot{\psi}\sin\gamma.\end{aligned}\right\} \tag{4.12}$$

When performing a 360° coordinated banked turn in the horizontal plane (when the sidesliding angle is zero), the expressions (4.12) take the form:

$$\left.\begin{aligned}\overline{\omega}_{x1} &= -\tfrac{\rho Sl}{4m}\tfrac{\vartheta}{\cos\gamma}(c_z^{\delta_{\rm H}}\delta_{\rm H} + c_y\sin\gamma\cos\gamma),\\ \overline{\omega}_{y1} &= -\tfrac{\rho Sl}{4m}(c_z^{\delta_{\rm H}}\delta_{\rm H} + c_y\sin\gamma\cos\gamma),\\ \overline{\omega}_{z1} &= \tfrac{\rho Sb_{\rm a}}{2m}(c_z^{\delta_{\rm H}}\delta_{\rm H} + c_y\sin\gamma\cos\gamma){\rm tg}\gamma.\end{aligned}\right\}, \tag{4.13}$$

where $b_{\rm a}$ is the mean aerodynamic wing chord.

Expressions (4.13) determine the components of the angular velocity along the axes of the coordinate system associated with the aircraft; with the help of these expressions, the total angular velocity of the aircraft performing a 360° coordinated banked turn can be found. Moving from unitless angular velocities to dimensional ones, squaring the components and adding these squares, we obtain the square of the total angular velocity and then the total angular velocity. Thus, we find:

$$\omega = -\frac{g}{V}\cdot\frac{m_y^{\delta_{\rm H}}}{m_y^{\delta_{\rm H}} - \frac{\rho Sl}{4m}\left(\frac{\vartheta}{\cos\gamma}m_y^{\overline{\omega}_x} + m_y^{\overline{\omega}_y}\right)c_z^{\delta_{\rm H}}}{\rm tg}\gamma, \tag{4.14}$$

where g is the acceleration of gravity.

Assuming the forces acting on the aircraft do not depend on the position of the rudders $\left(c_z^{\delta_{\rm Rudder}} = 0\right)$, we obtain an even simpler expression:

$$\omega = -\frac{g}{V}{\rm tg}\gamma. \tag{4.15}$$

Taking into account that $V/R = \omega$, we obtain, considering the above assumptions, the simplest expression for calculating the radius of a 360° coordinated banked turn R performed in the horizontal plane:

$$\mathrm{R} = \frac{V^2}{g\,{\rm tg}\gamma}. \tag{4.16}$$

In the general case, the radii of curvature of the trajectories for the right and left turns differ, for example, taking into account the asymmetry of the aircraft, which occurs when one of the engines fails. When performing a turn in wind conditions with a constant roll, the trajectory of the right turn and the left turn is not circles, and this further complicates the calculation of the ST boundaries.

If in the considered case, in expression (4.16), the maximum allowable bank angle (for reasons of allowable lateral overload), $\gamma_{\rm all.}$, is used, then, with the use of the current value of the speed $\boldsymbol{V}$, the current minimum permissible turn radius can be calculated, and with the predicted speed value, the predicted minimum permissible turn radius can be obtained.

The method considered above and its variants make it possible for the crew to properly estimate the degree of hazard and to make the right decision about the need and nature of maneuvering *before* a hazardous situation evolves.

One of the variants of the device implementing the claimed method is presented in Fig. 4.17 as a block diagram. The operation of the device presented in Fig. 4.17 is described in detail in [1, 2]. The crew shall analyze the current dynamic capabilities of the aircraft for maneuvering; thus it is useful to display a flown segment of the trajectory on the display as limited by the selected scale of the screen; to this end, an additional block of coordinates of the flown trajectory is added to the device (see Fig. 4.17).

Figure 4.17 contains the following designations: 1—navigation system, 2—obstacle detector, 3—signaling device, 4—video generator, 5—display, 6—calculator of the current dynamic state parameters, 7—coordinate calculator, 8—predicted path calculator, 9—comparator, 10—aeronautical information database, 11—control unit, 12—terrain relief database, 13—calculator of the minimum allowable turn radius, 14—calculator of predicted minimum allowable turn radius, 15—generator of the safety tunnel, 16—scanning group of the comparator, and 17—memory unit for the flown trajectory.

Thus, the method discussed above makes it possible for the aircraft crew to properly estimate the degree of hazard and to make the correct decision about the need and nature of maneuvering.

The proposed method allows for the implementation of new CAS advantages:

- notify of a hazardous proximity with the terrain, in good time sufficient to select an avoidance maneuver, including a reverse turn, without exceeding the permissible overloads;
- increase the crew's awareness of the terrain that poses a threat by generating notifying information on the hazardous terrain in the area of the intended maneuvering and thereby to simplify the decision on whether to perform a maneuver and to select a maneuver for avoiding the potentially hazardous terrain;
- simplify the crew's task of assessing the degree of hazard presented by the terrain ahead, and selecting the necessary maneuver by displaying the terrain projections, as well as flown and predicted paths in one scale;
- reduce the crew load regarding prevention of hazardous proximity with the terrain by timely notification of the hazardous terrain located in the area of intended maneuvering.

The implementation of these advantages allows for reduction the probability of a collision with the terrain and to expand the crew's capabilities to prevent hazardous proximity with the terrain by increasing the amount of information about risks of the geometry of the aircraft and the terrain, and also to increase the ergonomics of information presentation to the crew by displaying the combined projections of the terrain, as well as flown and predicted paths.

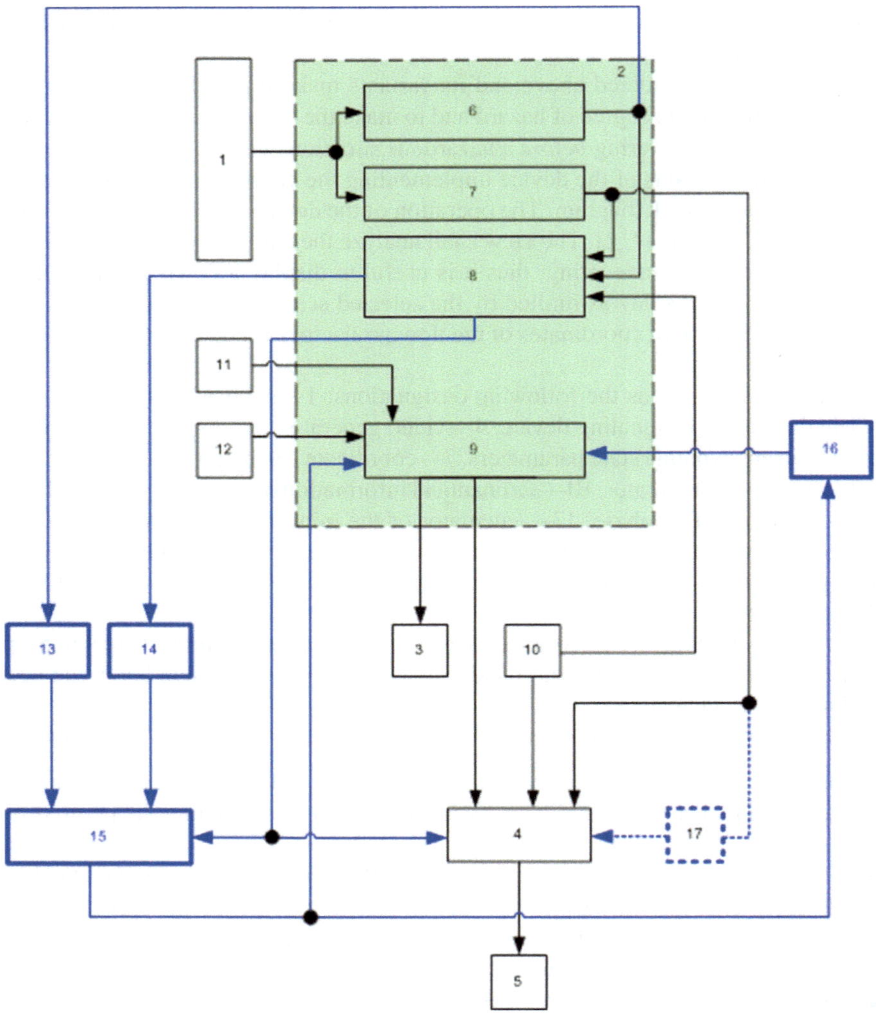

Fig. 4.17 Block diagram of the device implementing the method for preventing collisions with the underlying surface

Let us perform an expert evaluation of improving flight safety using the method considered above for determining hazardous terrain, taking into account the capability of a reverse turn.

Assume that in the direction of the flight there is an area such as a mountain canyon. The pilot must decide on the continuation of the flight along the chosen trajectory. This can be done by analyzing the nature of the underlying surface in the direction of the flight; so it is necessary to request and analyze the available information pages on the display with the corresponding projections of the underlying surface, and if there is a good visual range, to analyze the external environment also. The analysis domain is the safe corridor that is necessary for the continuation of the flight, taking into account the possibility of a reverse turn.

The results of experimental studies [16] show that the time for requesting the necessary additional information on the display is 6.5 s on the average, and the time required to select the necessary maneuver, taking into account a variety of factors and depending on the specific situation, can reach 20–30 s. It should be taken into account that in a stressful (emergency) situation, the probability of making an erroneous decision is high enough.

Continuous automatic comparison of the safety tunnel with the terrain data in the device included in the CAS, based on the algorithms proposed in the method under consideration, makes it possible to notify almost instantaneously of the presence of hazard in the direction of the flight and take the necessary measures to prevent the aircraft from ending up in an emergency situation.

Let us estimate the ergonomics factor $\mathbf{F}_{\mathbf{Erg}}$ as the ratio of the time taken to decide on the continuation of the flight along the current trajectory when using the proposed method for detecting a hazardous terrain and without using it. Considering the fact that to make such a decision, the pilot, through independent analysis, will need the following time at least: 3.5 s to switch his/her attention, $3 \cdot 6.5 = 19.5$ s to request help pages with additional information on the display screen, 2.7–14 s for the analysis of the external environment and at least 10 s to make a decision, then the total time will be 35.7 s (3.5 + 19.5 + 2.7 + 10).

The proposed method provides automatic notifications about the presence of a hazard with a display of its nature and a possible way out to avoid an emergency situation. The time spent by the pilot to decide on the nature of the maneuver in this case can be estimated as follows: 3.5 s for switching his/her attention, 2.7–14 s (for definiteness, we take the average value of 8.3 s) for a visual analysis of the external environment, up to 10 s on the decision on the maneuver. Total: 21.8 s (3.5 + 8.3 + 10).

The CAS, in which the proposed method for determining a hazardous terrain is implemented, is characterized by the ergonomics factor $\mathbf{F}_{\mathbf{Erg}} = 35.7/21.8 \approx 1.6$.

To estimate the reduction in the probability of an aircraft accident, we estimate the additional distance to the obstacle that the pilot of the aircraft will gain when deciding whether to perform an avoidance maneuver after receiving a CAS notification or when making a decision independently, but 13.9 s (35.7 − 1.8) later. Let us assume that the ground speed of the aircraft is $V = 200$ m/s.

Thus, if there is a CAS onboard the aircraft that implements the method considered above, we obtain the distance traveled in the direction of the obstacle until the decision on the maneuver is taken: 200·21.8 = 4360 m, and in the absence of notification, we obtain the distance 200·35.7 = 7140 m. Hence, we get the estimate: $\mathbf{F}_{\mathbf{Erg}} = 7140/4360 \approx 1.6$.

The above method of generating notifications on a hazard by comparing the safety tunnel with the terrain makes it possible for the crew to significantly shorten the time for making a decision on maneuvers to ensure a safe continuation of the flight. The expert estimate shows that the proposed method allows saving up to 14 s for the pilot to make a decision on maneuvering, which gives an additional margin in the distance to the obstacle as compared to the option of not using the proposed method. Obviously, flight safety does increase. It should also be noted that these estimates do not take into account possible erroneous decisions of the pilot, especially in a stressful situation. The proposed method also reduces the probability of erroneous pilot decisions that can lead to a catastrophe.

Next, consider another method, which is based on the use of the algorithms described above, but also provide further improvement of flight safety for low-altitude flights under the conditions of hazardous terrain.

4.4 Method for Improving Flight Efficiency and Safety by Analyzing the Space Inside a Corridor Safe for Flight

This method is based on the definition of safe trajectories of the aircraft moving over a hazardous terrain and can be used for aircraft of various types, including those that have the capability, like helicopters, to take off vertically and hover—gyroplanes, autogyros, etc. A detailed description of this method is given in [2, 3, 7].

For all known ways to improve flight safety [4, 5, 11, 12, 25, 26], there is an improper probability of preventing aircraft from colliding with the terrain, since they do not consider situations when, if a hazardous terrain is detected along the trajectory and it is impossible to circumvent it by climbing, the aircraft will have to perform a reverse turn, which may not be possible when flying at low altitudes in areas with mountainous terrain, during a landing or takeoff from a mountain airfield, or when flying in canyons, etc. The helicopter option to reduce the ground speed down to zero (hovering) will increase the mission time, and this is often unacceptable.

The use of method [6] is limited by the fact that under conditions when a notification is received of the intersection of the safety tunnel (ST) with the terrain, the pilot must independently assess the degree of hazard and determine the necessity and nature of maneuvering to avoid a hazardous situation. This leads to a significant increase in the psychophysical load of the pilot and may lead to an erroneous decision.

The objective of the method in question is to improve flight safety by analyzing the space inside a safe corridor for the flight and, if necessary, outside it, and to provide recommendations to the pilot regarding possible actions to avoid the hazardous situation. The terrain intersecting with the ST will be defined as potentially hazardous.

The idea of the claimed method for preventing collisions of the aircraft with the terrain is as follows: An ST is formed around the predicted path of the aircraft movement, a potentially hazardous terrain is determined, and then a recommendation is generated and displayed in a form convenient for the pilot that provides advice on the most effective continuation of the flight under the existing conditions.

The sequence of operations for implementing the claimed method is shown in Fig. 4.18. The plan-view projections for the ST, the predicted path, the protection space, and the alarm/warning signal region are shown in Fig. 4.19a, and the corresponding cross projections are depicted in Fig. 4.19b.

Let us consider how the proposed method for improving flight safety is implemented. To this end, we return to Fig. 4.18, from which it follows that using the navigation system (GNSS) onboard the aircraft, the position of the aircraft and the parameters of the current aircraft dynamic state are determined (ground speed W_{gnd}, vertical speed W_y, track angle TA, turn rate ω_y, etc.); these are used to extrapolate the position of the aircraft over a preset time interval (T_{fl}) and to calculate the predicted path (see block of algorithms 3).

Further, in block 4 (see Fig. 4.18), the minimum permissible turn radii are calculated (both to the right and to the left of the predicted path) on the basis of the dynamic state parameters, and in block 5, the predicted (at the time when the hazardous terrain was detected) turn radii are calculated based on the calculated predicted dynamic parameters of the aircraft.

After this, in block 6, the ST boundaries are determined, within which the predicted path must be located.

The ST boundaries are determined taking into account the aircraft capability to perform a reverse turn and in accordance with the predicted minimum permissible turn radii. In this case, the ST is compared to the terrain in block 7 (within the preformed sample from the terrain relief database obtained from block 8) by scanning the space within the ST.

In the case some part of the terrain crosses the ST, a notification about the presence of a potentially hazardous terrain is generated (see block 10 in Fig. 4.18). If a potentially hazardous terrain is detected, block 11 generates a protection space with an alarm/warning signal region, and block 12 predicts the alarm/warning region for the ST length. Then, both in the case with the ST and the terrain intersection on either side of the predicted trajectory, and in the case of the terrain intersection on one side of the predicted trajectory with simultaneously predicted alarm/warning signal region, the space is explored by varying the parameters of the safe corridor and the predicted trajectory (see block 15).

On the basis of this exploration, safe variants of flight parameters and the corresponding maneuvers of aircraft are defined. The aircraft maneuvers corresponding to safe variations of the flight parameters are displayed in a form convenient for the pilot, using the appropriate algorithms in block 16.

The ST in Fig. 4.19 is presented, respectively, in the form of a plan-view (a) and cross (b) projections of the space domain, which is directly connected with the current aircraft position and determined (in shape and orientation in space) by the predicted trajectory.

Dependence of the ST parameters on the parameters of the aircraft movement (current and predicted minimum permissible turn radii, maximum error in the aircraft position, minimum safe side distance between the aircraft and the terrain, ground speed, vertical speed, track angle, angular turn speed, minimum allowable flight

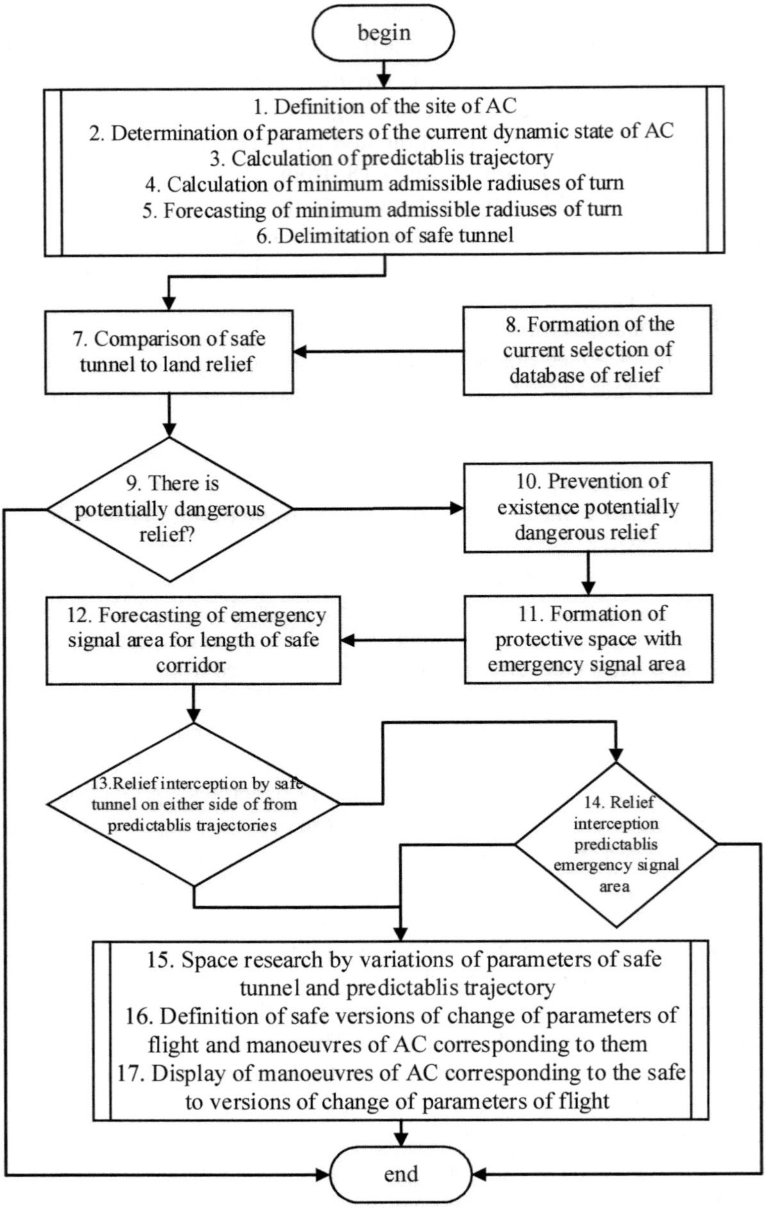

Fig. 4.18 Sequences of operations for implementing the method for determining safe Trajectories

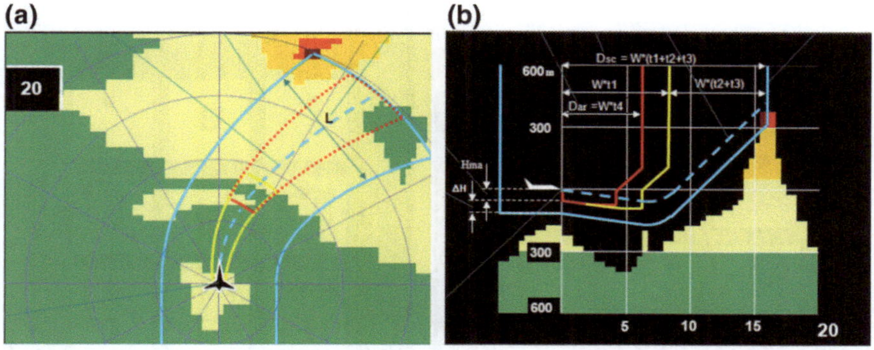

Fig. 4.19 Plan-view (**a**) and cross (**b**) projections of the safety tunnel

altitude H_{ma} and additional height margin ΔH) was described in detail in the previous section of this paper.

The length of the ST part located in the direction of the flight (D_{ST} in Fig. 4.19b) is defined as the product of the aircraft ground speed and the flight time to the ST front edge. The flight time is the sum of three characteristic time intervals:

$$D_{\text{ST}} = W_{\text{gnd}} \cdot (t1 + t2 + t3), \tag{4.17}$$

where *W* is the ground speed of the aircraft, $t1$ is the maximum possible flight time to the obstacle that may trigger an alert/warning signal (the length of the alert/warning generation area), $t2$ is the time required for the pilot to decide on the need and nature of additional maneuvering to avoid crossing the alert/warning region with the terrain, $t3$ is time required to change the current values of the flight parameters to the values corresponding to the maneuver selected to flyby the hazardous terrain.

The length of the alert/warning signal region (D_{ar} in Fig. 4.19) is determined in accordance with expression:

$$D_{\text{ar}} = W_{\text{gnd}} \cdot \text{t4}, \tag{4.18}$$

where $t4$ is the maximum possible flight time to an obstacle that may trigger an alert/warning (length of the alert/warning signal region).

The configuration of the ST boundaries is converted with a rate needed to update the information, and thus a BC with a width *L* (see Fig. 4.19a) is adapted to the current aircraft dynamic status and to the flight phase and mode.

If a potentially hazardous terrain is detected, a corresponding notification is generated and in the cases mentioned above, which require exploration of the space and definition of safe trajectories and the aircraft maneuvers necessary for their implementation; the above maneuvers are displayed, for example, in the form shown in Fig. 4.20.

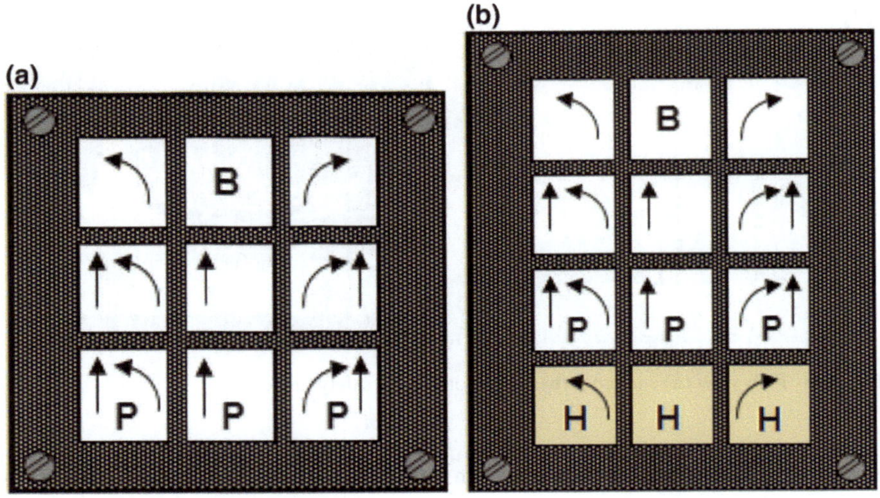

Fig. 4.20 Displaying recommended maneuvers: **a** for aircraft, **b** for helicopters

The example of displaying the recommended maneuvers shown in Fig. 4.20a involves displaying information for the pilot of the aircraft in such a way that each element corresponds to one of the possible maneuvering options for avoiding a potentially hazardous terrain. On the elements of the left column, maneuvers with a left turn are displayed (indicated on the panel element by an arc with an arrow indicating the counterclockwise direction of the turn), maneuvers with straight movement are displayed on the elements of the central column of the panel, and maneuvers with the right turn (indicated on the panel element by an arc with an arrow indicating the clockwise direction of the turn). In this case, maneuvers performed without braking are displayed on the first and third elements of the first line of the matrix panel, and those with braking on the second element (indicated by "B"). The elements of the second line show maneuvers performed with climbing (indicated by an arrow pointing upwards) and braking, and elements of the third line display maneuvers with climbing and an increase in engine power (denoted by "P").

The pilot may select the maneuver that is most suitable for accomplishing the mission task to be solved, taking into account the current conditions for its solution. This approach substantially facilitates the pilot's task of making a decision in the current situation, since it allows evaluation of the degree of hazardous proximity with the terrain before a hazardous situation evolves, does not require preventive viewing of the terrain display, simplifies the search for safe maneuvers to avoid a potentially hazardous terrain, and reduces the crew's psychological load. It is possible to avoid the evolvement of a hazardous situation onboard the aircraft with timely maneuvering performed on the basis of the generated signal about a potentially hazardous terrain and using the indication of possible safe maneuvers to avoid a potentially hazardous terrain.

An example of displaying the recommended maneuvers shown in Fig. 4.20b involves displaying information for the pilot of the helicopter. Unlike the panel shown in Fig. 4.20a, the panel shown in Fig. 4.20b has an additional line for displaying the recommended maneuvers, which end with hovering of the aircraft (indicated by "H").

The degree of the terrain hazard in the predicted flight is analyzed from the moment of receiving the notification signal caused by the contact of the ST with the terrain according to the TDB. The ST parameters are determined by the parameters of the aircraft movement both current and predicted for a fixed flight time. If a lateral part of the ST contacts with the terrain (left or right one or both at once) and the analysis of the relative terrain and ST geometry shows that the alert signal about the hazardous terrain will not change to a warning signal when the flight is continued, then the terrain hazard is considered insignificant and no alternative trajectories to flyby the hazardous terrain are calculated. Otherwise, a control signal is generated, with which the recommended trajectories are calculated to avoid the hazardous terrain.

The hazard for the flight along each of the calculated trajectories is estimated by the same criterion as the hazard for the predicted flight (by comparing the terrain with the ST of each of the calculated trajectories).

The number and type of calculated trajectories to avoid a hazardous terrain are determined by the following factors:

- aircraft type (helicopters have more types because of the greater number of possible options for changing the movement parameters);
- technical condition of the aircraft (engine failure, malfunction in the control system, etc.);
- fuel onboard (with a significant stock of fuel the large weight of the aircraft can degrade the climb performance and with a small stock it is impossible to increase the engine power);
- nature of the surrounding terrain (with complex terrain, the capabilities for changing flight parameters can be significantly limited);
- flight conditions (flight altitude, temperature outside the aircraft, precipitation, wind, etc.).

The calculated trajectories involve changing one or more flight parameters to accomplish the task of avoiding the hazardous terrain. The necessary changes in the flight parameters shall be made taking into account the prompt return to the main flight program (en route flight, monitoring the terrain, conducting combat operations, etc.).

If there are more than three possible trajectories to avoid the hazardous terrain, it is advisable to select three most preferable ones and display them to the pilot on a special panel. This will facilitate the problem of selecting a maneuver for the pilot.

If the design of the panel for indicating the recommended maneuvering to void the hazardous terrain will ensure displaying each maneuver on one spot of the panel anytime, it will be useful to ensure correct perception of the maneuver by the pilot. However, this will require a large panel (because of the wide variety of possible recommendations); and this, given the limited number of recommendations displayed (three), will lead to non-involvement of the main part of the panel. Therefore, a three-position display can be used that shows the maneuvers assigned to them in groups formed according to general criteria.

In the proposed method, an alarm about the hazardous terrain is generated and if necessary, recommendations are given for maneuvering to avoid the hazardous terrain. At the same time, it is always possible for the pilot to change the flight path, different from the recommended ones and depending on factors not taken into account in analyzing the hazard of the calculated trajectories (aircraft damage, thunder, etc.).

The structural diagram of a device that can improve flight safety by applying the method under consideration is shown in Fig. 4.21. In the presented device, algorithms and calculation methods described above are fully implemented.

The operation of the presented device is described in detail in [7].

The proposed method and algorithms, techniques, and devices used in its application ensure implementation of new CAS advantages:

- notify of hazardous proximity with the terrain for a time sufficient to select the recommended avoidance maneuver, including a reverse turn, without exceeding the permissible overloads, including the hovering maneuver for aircraft capable of it;
- increase the crew's awareness of the terrain hazard degree by displaying or not displaying the recommended maneuvers to avoid the potentially hazardous terrain and thereby simplify the task of deciding whether to maneuver to avoid a potentially hazardous terrain that caused the notification signal;
- simplify the crew's task of selecting the necessary maneuver to avoid the potentially hazardous terrain by displaying recommended maneuvers;
- reduce the crew load in preventing hazard proximity to the terrain by timely notifying of a potentially hazardous terrain and displaying a message about the need for maneuvering and recommended maneuvers (if necessary).

Implementation of the above-mentioned advantages ensures reduction of the probability of accidents through more detailed analysis of the risks of the geometry of the aircraft and the terrain, to simplify the pilot's task of determining the need and nature of maneuvering, and to improve the ergonomics of information presentation to the crew by displaying the recommended maneuvers.

Let us perform an expert evaluation of improving flight safety of using the above method for analyzing the space inside a corridor safe for the flight and displaying the recommended maneuvers.

As before, assume that in the direction of the flight there is an area such as a mountain canyon. The pilot must decide on the continuation of the flight along the chosen trajectory. If there is no CAS onboard that uses the method discussed above, the pilot may need 35.7 s (see Sect. 4.3).

Comparison of the safety tunnel with the terrain relief on the basis of the proposed method makes it possible to receive a notification about the presence of a hazard in the direction of the flight and recommendations for implementation of a safe maneuver.

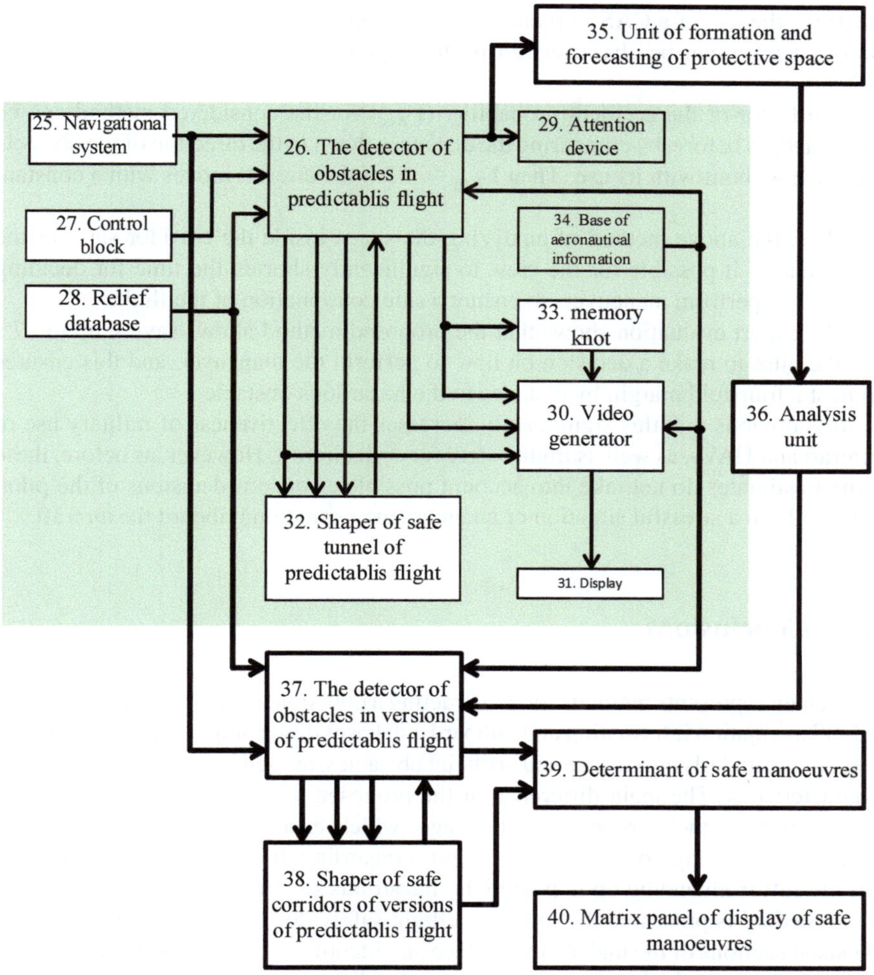

Fig. 4.21 Structural diagram of a device implementing the method for analyzing the space inside the safe tunnel

Let us estimate the ergonomics factor $\mathbf{F}_{\mathbf{Erg}}$ as the ratio of the time taken to decide on the continuation of the flight along the current trajectory when using the proposed method for detecting a hazardous terrain and without using it.

When a notification of the hazard is given, the time taken by the pilot to make a decision will be as follows: 3.5 s for switching his/her attention and, possibly, 5–10 s for a visual analysis of the external environment in order to determine whether the proposed type of maneuver can be performed. In this case, if there is a recommendation, (see Fig. 4.20), this will take no more than 5 s. Total: 8.5 s.

Thus, the use of a CAS implementing the proposed method for detecting a hazardous terrain can be characterized by the ergonomics factor $\mathbf{F_{Erg}} = 35.7/8.5 \approx 4.2$.

The factor of the accident probability ($\mathbf{F_{Acc}}$) for the considered method can be estimated, as before, by comparing the distances flown in the direction of the obstacle (hazard) without/with its use. Then $\mathbf{F_{Erg}} \approx 4.2$ if the aircraft moves with a constant speed.

Thus, the above method of analyzing the space inside the corridor safe for the flight makes it possible for the crew to significantly shorten the time for deciding whether to perform maneuvers to ensure a safe continuation of the flight.

The expert evaluation shows that the proposed method allows saving up to 27 s for the pilot to make a decision on how to perform the maneuver, and this ensures almost a four-fold margin by distance to the hazardous obstacle.

It is obvious that this significantly increases the effectiveness of military use of aircraft and UAVs, as well as flight safety for civil aircraft. However, as before, these expert estimates do not take into account possible erroneous decisions of the pilot, especially in a stressful situation or an emergency situation onboard the aircraft.

4.5 Conclusions

This chapter presents methods for constructing a collision avoidance system with the GNSS navigation information and onboard databases describing the terrain (topography) of the underlying surface and artificial obstacles, as well as aircraft performance characteristics. The main difference of the proposed methods from the previously known ones is their ergonomic advantage, which ensures improvement of flight safety by providing notifications to the pilot regarding the possibility of preventing the aircraft from ending up in potentially hazardous situations.

A method is presented that improves flight safety due to three-dimensional synthesis of sections of the underlying surface and identification of hazardous elements. The use of this method provides the pilot with additional time of about 15 s for flights under the conditions of a hazardous underlying surface and an additional height of up to 300 m. The proposed method is characterized by the ergonomics factor $\mathbf{F_{Erg}} = 2.6$ and a reduction in the accident factor $\mathbf{F_{Acc}} = 1.6$.

A method is also presented for determining the direction of a turn to prevent aircraft collisions with the underlying surface based on the analysis of the nature of the underlying surface in the direction of the flight, taking into account the aircraft performance characteristics. The method provides the pilot with additional time of about 33 s for flights under the conditions of a hazardous underlying surface and an additional distance to an obstacle of about 6.6 km. This method is characterized by the following factors: $\mathbf{F_{Erg}} = 3.5$, $\mathbf{F_{Acc}} = 3.5$.

A method is described for detecting a hazardous terrain, taking into account the possibility of a reverse turn, in which the minimum permissible radii of the turn are calculated on the basis of the dynamic equations of movement, which provides advance notifications for the pilot of the fact that the flight in the selected direction represents a potential hazard if a reverse turn is needed. The use of this method makes it possible to improve flight safety due to additional time of about 14 s and an additional distance to an obstacle of about 2.9 and 6.6 km. This method is characterized by the following factors: $\mathbf{F}_{\mathbf{Erg}} = 1.6$, $\mathbf{F}_{\mathbf{Acc}} = 1.6$. The availability of notifications formed in the proposed method makes it possible to abandon the flight along the planned trajectory and completely exclude the probability of an accident.

A method is also presented for analyzing the space inside a safe corridor, which can be used not only for aircraft, but also for helicopters. In this method, the display means form recommendations for the pilot on maneuvering to improve flight safety. Using this method ensures the improvement of flight safety due to additional time of about 27 s. This method is characterized by the following factors: $\mathbf{F}_{\mathbf{Erg}} = 4.2$, $\mathbf{F}_{\mathbf{Acc}} = 4.2$.

The next chapter describes integrated technical solutions for the joint SLS and CAS use that address a number of fundamental problems in improving flight safety.

References

1. Baburov VI, Galperin TB, Maslov AV, Sauta OI (2009) Method for preventing an aircraft collision with a mountainous relief with the capability of a reverse turn. Radioelectronics issues. Series RLT-2009-Issue 2, pp 114–126 (in Russian)
2. Baburov VI, Galperin TB, Maslov AV, Sauta OI (2010) On the development of the ideology of methods for preventing aircraft collisions with terrain. Radioelectronics issues. Series RLT-2010-Issue 2, pp 180–190 (in Russian)
3. Baburov VI, Volchok YuG, Gubkin SV, Maslov AV, Sauta OI, Galperin TB, Pukhov GG, Moroz NV, Vodov MA (2005) Method for preventing an aircraft collision with ground and a device based thereon. Patent No. RU 2 262 746, publ. on 20.10.2005, Bul. No. 29 (in Russian)
4. Baburov VI, Mkhitaryan VA, Kontsevich LV, Sauta OI, Galperin TB (2006) Collision prevention system. Patent No. RU 2 271 039, publ. on 27.02.2006, Bul. № 6 (in Russian)
5. Baburov VI, Volchok YuG, Galperin TB, Gubkin SV, Maslov AV, Sauta OI (2007) A Method for preventing an aircraft collision with a terrain relief and a device based thereon. Patent No. RU 2 301 456, publ. on 06.20.2007, Bul. No. 17 (in Russian)
6. Baburov VI, Volchok YuG, Galperin TB, Gubkin SV, Maslov AV, Sauta OI (2009) Method for preventing aircraft and helicopter collisions of with terrain and a device based thereon. Patent No. RU 2 376 645, publ. on 20.12.2009, Bul. No. 35 (in Russian)
7. Baburov VI, Volchok YuG, Galperin TB, Gubkin SV, Maslov AV, Sauta OI (2011) Method of notification of the aircraft location relative to the runways during the approach. Patent No. RU 2 410 753, publ. on 01/27/2011, Bul. No. 3 (in Russian)
8. Baburov VI, Volchok YuG, Galperin TB, Gubkin SV, Maslov AV, Sauta OI, Rogova AA (2011) Method of notification of the aircraft location relative to the runways during the approach and roll-on operation. Positive decision on granting a patent following the application for an invention No. 2011113706/11 (020353), priority of 04/04/2011 (in Russian)
9. Gomin SV, Makelnikov AA, Maslov AV, Priyemysheva AA, Sauta OI, Semenov GO, Sobolev SP (2005) Program for warning and emergency alarms generation. Certificate of official registration of computer programs. RF №2005611918. Federal Service for Intellectual Property, Patents and Trademarks (in Russian)
10. Frence patent No. 2731824, cl. G08G 5/04, application of 17.03.1995, publ. on 09/20/1998

11. France patent No. 2747492, cl. G08G 5/04, application of 15.04. 1996, publ. on 17.10.1997 (in French)
12. France patent No. 2 773 609, cl. G01C 5/00, application of 12.01.1998, publ. on 16.07.1999 (in French)
13. US Patent No. 5 892 462, cl. G08G 5/04, application of 20.06.1997, publ. on 06.04.1999
14. US Patent No. 6 021 374, cl. 163/00, application of 09.10.1997, publ. on 01.02.2000
15. US Patent No. 6 480 120, cl. G08B 23/00, application of 15.04.1996, publ. on 12.11.2002
16. Minimum Performance Standards—airborne ground proximity warning equipment. RTCA DO-161A, 1976
17. Russian patent No. 2 211 489, cl. G08G 5/04, application of 11.01.1999, publ. on 27.08.2003 (in Russian)
18. France patent No. 2 731 824, cl. G08G 5/04, application of 17.03.1995, publ. on 20.09.1998 (in French)
19. France patent No. 2 747 492, cl. G08G 5/04, application of 15.04.1996, publ. on 17.10.1997 (in French)
20. Kublanov MS (2000) Aerodynamics and flight dynamics. Moscow State University, Moscow (in Russian)
21. Ostoslavsky IV, Strazheva IV (1965) Flight dynamics. Stability and controllability of aircraft. Mechanical Engineering, Moscow (in Russian)
22. Ostoslavsky IV, Strazheva IV (1969) Flight dynamics. Aircraft trajectories. Mechanical Engineering, Moscow (in Russian)
23. Russian Patent No. 2 312 787, cl. B64D 45/04, application of 25.11.2004, publ. on 20.12.2007 (in Russian)
24. Abstract of a study conducted by the European Aviation Safety Agency (EASA) for 2009. Flight safety issues, No. 11, 2010. VNIITI, Moscow (in Russian)
25. US application No. 2006290531, cl. G08G5/54, publ. on 28.12.2006
26. Sauta OI (1991) Substantiation of the applicability of SRNS navigation radio beacons to support aircraft approaches (in Russian)

Chapter 5
Integrated Technical Solutions on the Joint Use of Technologies Applicable in Collision Avoidance Systems and Satellite-Based Landing Systems

This chapter describes the basic principles of constructing an integrated flight safety system on the basis of the satellite-based landing system (SLS) and the collision avoidance system (CAS) using GNSS technologies. The proposed technical solutions are partially described in [1–3].

The principles considered in this chapter are based on the assumption that both CAS and SLS systems are installed onboard the aircraft. Currently, all civil aircraft are mandatorily equipped with a CAS. At the same time, the SLS development has gained momentum only in recent years, and the peak of its implementation in Russia is expected in a few years when the appropriate airborne equipment will be installed on the majority of aircraft. It should be noted that in Russia, airfields have been equipped with ground SLS subsystems (LAASs) at a rapid pace and by 2016 about 100 airfields will have already been equipped with LAASs.

5.1 Principles of Constructing an Integrated Flight Safety Enhancement System Based on the Collision Avoidance System and the Satellite-Based Landing System

As was shown in Chap. 2 (see Fig. 2.1), at the III stage of the process of SLS and CAS building, prospective elements for integration and complexing shall be identified. It was also noted there that this procedure includes consideration of the functional, hardware, and software elements of each of the systems.

The structures for SLS and CAS building were discussed in detail in Chap. 2. An SLS feature is the availability of ground and onboard subsystems, and the CAS is a purely onboard system.

Figure 5.1 shows the main SLS and CAS functional elements and indicates possible mutual relationships between them regarding the use of the same type of data or formation of functionally similar parameters.

© Springer Nature Singapore Pte Ltd. 2020
Baburov S.V. et al., *Development of Navigation Technology for Flight Safety*,
Springer Aerospace Technology, https://doi.org/10.1007/978-981-13-8375-5_5

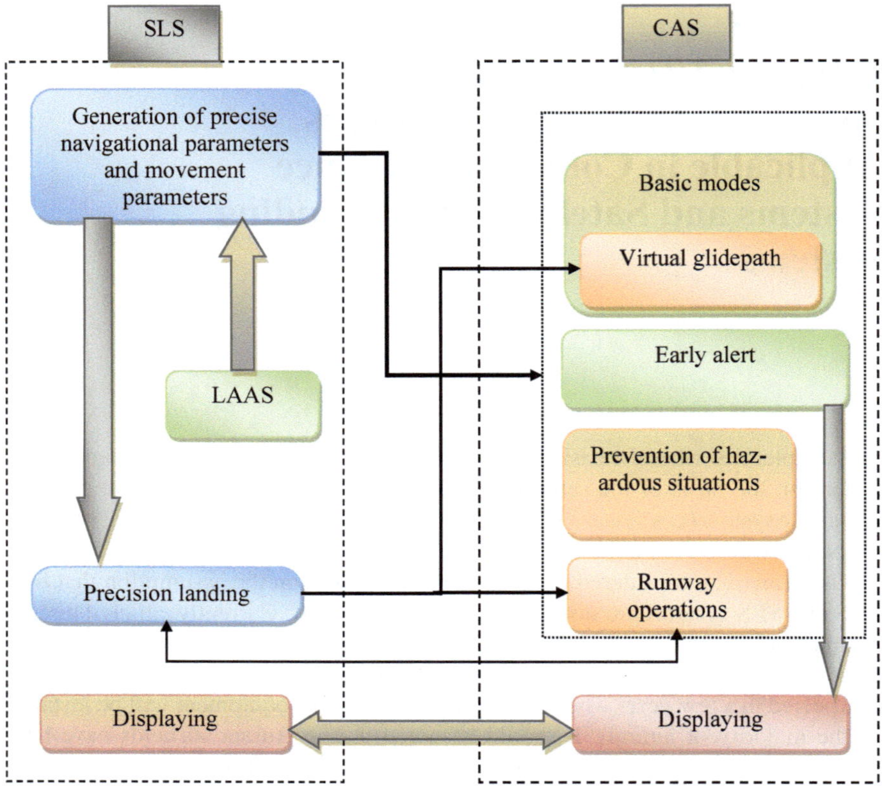

Fig. 5.1 Interrelationship of the SLS and CAS functional elements

The analysis of the relationships presented in Fig. 5.1 allows for the following conclusions:

1. A single source of navigation information, a single GNSS receiver, can be used to provide the efficiency of the SLS and CAS operation. Considering the fact that in accordance with the general requirements for landing systems [4–6], there is always a backup (redundancy) set of equipment in the "hot" standby mode in the SLS, it is advisable to use a GNSS receiver from the SLS equipment in the construction of an integrated system, as a source of information for the CAS.
2. The use of these aircraft deviations from the intended landing glide path in the SLS makes it possible to implement one of the basic CAS modes, mode 5 (excessive deviation from the landing glide path) in the absence of ILS-like instrument landing systems at the aerodrome. Moreover, this mode of operation makes it possible to modernize mode 5 for curved approaches, which is currently considered as one of the most promising areas for the development of landing systems in general.

3. In the basic CAS modes (see Sect. 2.3), the height (H_{ra}) data from the onboard radio-altimeters (RA) are used. Because of the high cost of the latter, RAs are not installed on small aircraft of general aviation. This generally reduces flight safety.

At the same time, it is with these aircraft that a significant number of accidents occur. The use of the GNSS data and the database of the underlying surface to generate the estimate of the height data H_{ra} of the aircraft above the Earth's surface support the operation of the basic CAS modes.

4. The CAS operating in basic mode 2 uses such parameter as the speed of closing between the aircraft with the underlying surface that is calculated on the basis of differentiating the radio-altimeter data and has errors that depend on the nature of the underlying surface over which the flight is performed. Experimental studies show that the error in determining the vertical velocity (V_y) when differentiating the radio-altimeter measurements can reach 5–10 m/s. Using the GNSS data on V_y values independent of the RA measurements and the onboard terrain relief database, it is possible to increase the accuracy of estimating V_y (GNSS measurements have high accuracy; the error in determining V_y with a probability of 0.95 is 0.1 m/s max), to significantly reduce the probability of false alarms in the CAS, and to improve the safety of low-altitude flights over a hazardous terrain and artificial obstacles.
5. Unconditional advantage of the integrated SLS and CAS use is to increase flight safety at the most dangerous flight phase, that is, during approaches and landings. An important task is to prevent aircraft from landing on an unauthorized runway, which can occur both in the case of an ATC controller error and due to pilot's fault. This also includes tasks of warning to the pilot about unacceptable deviations of the aircraft from the intended trajectory and preventing the aircraft from overrunning the runway after the landing. Lack of information and time for decision-making at the landing phase can lead to a catastrophe (accident).
6. The most important objective of improving flight safety is displaying both navigation information and the underlying surface in the terminal area, artificial obstacles in the runway area or some hindrance on the runway.

The above analysis shows that many of the functions performed by the SLS and CAS are complementary. This allows for their effective integration.

The interrelationship of the main SLS and CAS functional elements allows the definition of the following principles for the construction of an integrated safety system based on the integrated use of these systems. These principles include the following:

1. a unified approach to the selection of GNSS receivers used in the composition of onboard navigation and landing complexes and ground equipment ensuring additional advantages in the joint operation of the SLS and CAS equipment;
2. the use of unified navigation and landing information both for the purpose of generating various types of alerts/warnings in the CAS and for piloting during the SLS-based landing;

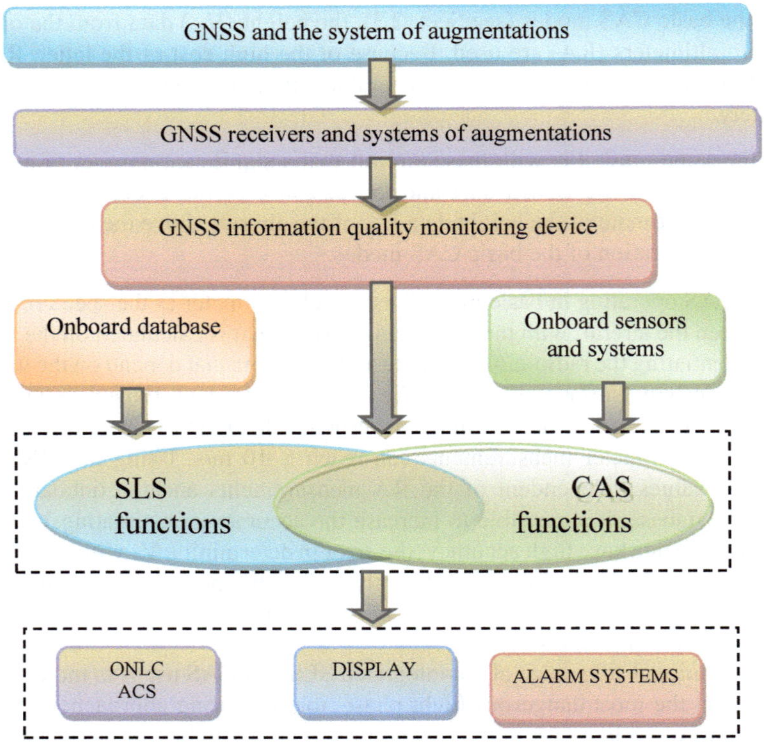

Fig. 5.2 Generalized structure of the integrated flight safety enhancement system based on the SLS and CAS

3. displaying of navigation and landing information with the information on the underlying surface and recommendations to the pilot on the implementation of safe maneuvers.

The use of the above principles determines the structure of an integrated SLS- and CAS-based flight safety system presented in Fig. 5.2.

The structure of the integrated flight safety enhancement system presented in Fig. 5.2 includes SLS and CAS functional elements based on the use of the GNSS. The GNSS is the unifying core of such integration. Flight safety for the integrated system is ensured by the virtual redundancy of the functions of one system by the other one, as well as by the hardware redundancy of the onboard GNSS modules, which is mandatory for the SLS.

Next, we will consider methods to improve flight safety during approaches and runway operations, which can be implemented using the principles of building integrated systems.

5.2 Method for Preventing Aircraft Landings on an Unauthorized Runway by Calculating a Virtual Glide Path

The proposed method is designed for use in airports with high traffic intensity and a large number of runways.

The method is based on the principle of forming protection spaces for the nearest runways and selecting the one on which the approach is to be performed. The selection is based on the current aircraft condition parameters. As a result of the selection, alerts are generated for the pilot regarding the runway that is currently most likely to be involved in an approach operation. Figure 5.3 shows an illustration explaining the operation of the runway selection algorithm in the proposed method.

At present, the method and equipment are known for notifying the crew on the location of the aircraft relative to the runway, which implements the Runway Awareness and Advisory System (RAAS) function [7–9]. According to the RAAS function, an alert is generated in the final approach segment regarding the selection of the runway and the location of the aircraft relative to the runway. These notifications are generated when certain conditions are met:

- when the aircraft is at a distance of 0.9–5.5 km from the runway threshold;
- in the range of heights of the aircraft above the runway of 90–215 m excluding the range of ± 15 m relative to the height reported within the specified range;
- in a given range of descent angles for a given height range (for example, 2°–15° for heights less than 150 m, which corresponds to the possible range of approach angles for most types of aircraft [10]) and when the aircraft is in the horizontal angular sector with vertex at the intended landing point and a bisectrix coinciding with the runway centerline [10].

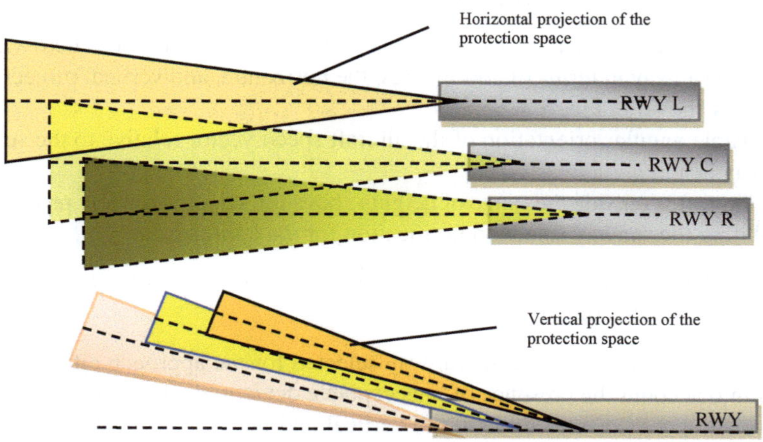

Fig. 5.3 Horizontal and vertical projections of protection spaces during the approach

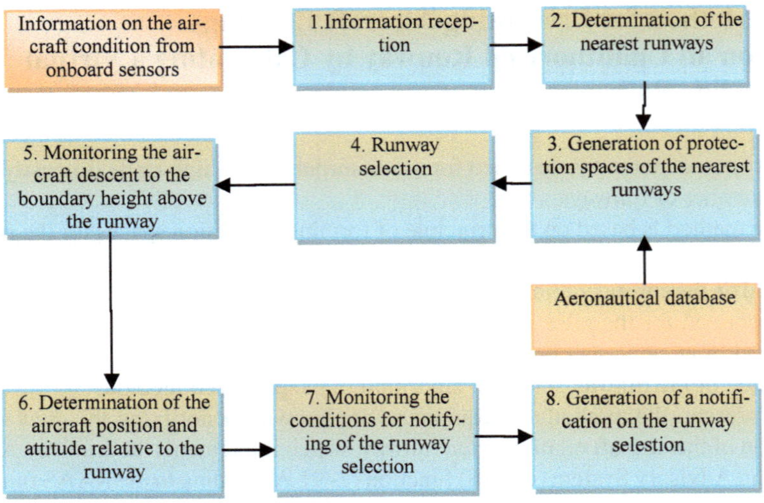

Fig. 5.4 Generalized sequence of operations in the method for preventing landings on an unauthorized runway

Various aspects of notifications of the location of the aircraft relative to the runway were considered in [7–9]. In [7], software issues are considered, in [8] issues of graphical representation of information to be displayed on the screen in the cockpit, including the potential collision of aircraft with external objects by displaying the intersection of the aircraft speed vectors and external objects. In [9], an extended set of issues is considered, including software and hardware implementation. The most common issue is the method [11]. The sequence of operations for implementing this method is shown in Fig. 5.4.

The above-mentioned conditions for notification of the runway selection are:

- The aircraft is inside a protection space, which is a region correlated with the selected runway in terms of coordinates, the horizontal, and vertical projection of which is shown in Fig. 5.3.
- Allowable angular orientation of the aircraft speed vector relative to the selected runway heading.
- The aircraft is outside the range of heights above the selected runway provided for inhibiting the runway selection notification for the approach.

The methods of the notification described in [7–9, 11] have a number of the following significant drawbacks:

- The notification (signal) of the runway selected for the approach either does not take into account the presence of unacceptable deviations of the aircraft from the intended approach trajectory (in the vertical and horizontal planes) or these are very inaccurate.

- When instrument landing systems are available and used in large airports, a notification of the above deviations of the aircraft on either side in a three-dimensional coordinate system is generated in the instrument landing system itself. However, if the instrument landing system fails, if there are failures in instrument landing system signals, or if there is no such system at the airport (this is true for most airports in Russia), the aircraft crew is not informed about the deviations of the aircraft position from the intended trajectory.
- The known methods either do not take into account the approach slope at all, or use the entire possible range of approach slopes typical for all aircraft used (from $2°$ to $15°$), while for each type of aircraft, the range of these slopes is considerably narrower. In addition, the mentioned range of slopes for each aircraft depends on the load, technical condition of the aircraft, and weather conditions.
- The generated notification of the runway selected for the landing is not refuted in any way even in the event of a subsequent violation of the notification conditions.

The task of the new proposed method is to improve the safety of the approach by forming a notification of the selected runway adapted to the type of aircraft and to generate notification about deviations of the aircraft position from the intended trajectory and deviations of the predicted landing point from the intended landing point.

The method in question is described in detail in [3]. The method involves the generation of a notification in which an alert/warning about the deviation of the aircraft from the intended trajectory is generated by displaying the deviation of the aircraft from the intended trajectory in the vertical plane and the deviation of the predicted landing point from the intended landing point in the runway plane.

Figure 5.5 shows a sequence of operations for the proposed method.

The sequence in Fig. 5.5, which describes the principle of the proposed method, includes the following operations: receiving information about the parameters of the aircraft dynamic state from its onboard sensors 1; determination of the nearest runway using the aeronautical database (AND) 2; generation of a protection space (PrSp) for the nearest runway 3 (see Fig. 5.3); selection of a runway for the approach 4; monitoring of the aircraft descent to the boundary height above each selected runway 5; determination of the aircraft position and attitude relative to the selected runways 6, above which the boundary height is not reached, when the aircraft sinks; monitoring the conditions for notifying of the runway selection 7; in addition to the above-mentioned boundary height, a zone of inhibition for the generation of the said alert/warning of the height is introduced that is defined by the type of aircraft and the need for other systems to generate a message saying that the aircraft has reached a certain height above the runway; generation of a notification 8 on selection of a runway in the event, these conditions are met.

If the selection of the runway for the landing is confirmed, the described operations are repeated until the aircraft sinks to the boundary height above each selected runway, below which no notification of the selection is generated. In the method under consideration, more stringent conditions are introduced as compared to those known to date, as well as conditions for generating notifications of elevations corre-

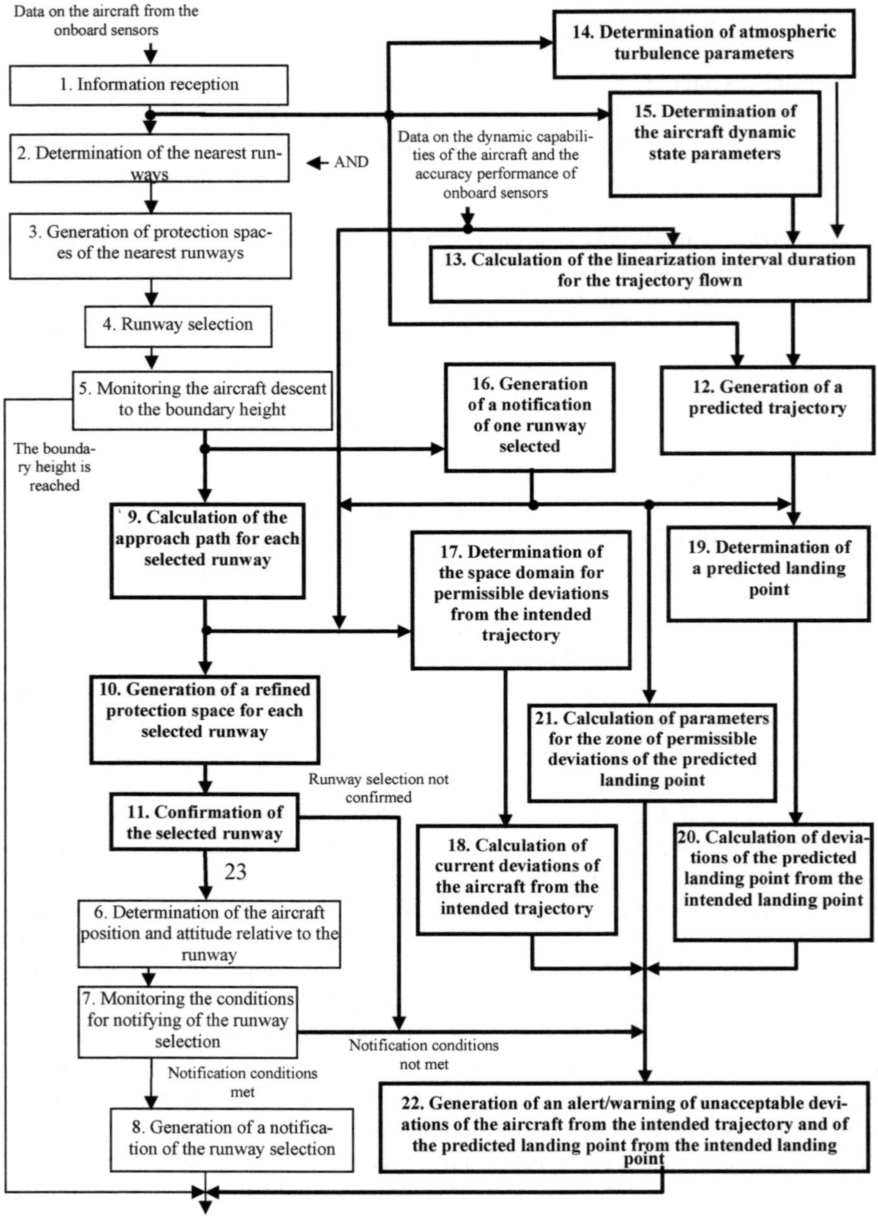

Fig. 5.5 Sequence of operations for the generation of runway selection notifications

sponding to the elevation angle values obtained on the basis of the allowable slopes for the intended approach trajectories for this type of aircraft.

Let us consider the most characteristic features of the proposed method.

The intended approach trajectory lies in a vertical plane passing through the runway centerline. Calculation of the intended trajectory 9 (see Fig. 5.5) is based on the onboard sensors' data on the height of the aircraft above each selected runway at a given distance from the runway threshold known for each type of aircraft and corresponding to a height of 200 m above the runway (about 4 km for a standard glide path). It is from this distance that, with visual approaches, a linear descent trajectory is usually maintained. The intended trajectory is calculated in the vertical plane passing through the runway centerline from the projection to this plane of the calculated point of passage of the given distance to the reference point, which is defined as the point above the middle point of the runway threshold (the point of intersection of the threshold with the runway centerline) at the intended height, usually about 15 m [12]. The height above the runway of said calculated point of passage the given distance is determined either as the current aircraft height at the time of passage (if the said distance is flown at a height differing from 200 m within a tolerance), or as the sum of 200 m and a half of the tolerance value with the appropriate sign. The intersection of the said linear trajectory with a horizontal plane passing through the middle point of the runway threshold is considered an intended landing point.

Refined PrSps for the selected runway 10 are generated by correcting the parameters of the vertical PrSp projections in accordance with an allowable range of approach slopes and the allowable range of deviations from this trajectory for this aircraft type.

The runway selection for the approach 11 is confirmed when the aircraft is in the refined PrSps of the previously selected runways.

It is after the runway selection is confirmed that the aircraft position and attitude described in [11] is determined with respect to the selected runway 5, the notification condition is monitored 7, and the notification is generated 8 (see Fig. 5.5).

If only one runway is selected after the generation of non-refined PrSps, then, after monitoring the aircraft descent to the boundary height above the selected runway 5, additional operations are performed, if necessary, to help the crew to land that consist of generation of notifications on deviations of the current position of the aircraft from the intended trajectory and deviations of the predicted landing point from the intended one.

To this end, the predicted trajectory is generated 12 after receiving information from the onboard sensors by the method of linear approximation of the flown trajectory.

For the basis of the approximation, the portion of the flown trajectory over a variable time interval is selected. The duration of the given time interval 13 is calculated up to the above approximation on the basis of the atmospheric turbulence parameters 14, and the parameters of the aircraft dynamic state 15 (speed vectors, acceleration, weight, etc.) determined from information of the onboard sensors, as well as data on the dynamic capabilities of the aircraft and the accuracy of the onboard sensors.

The aircraft dynamic capabilities are taken into account in such a way that the duration of the said time interval is calculated using calculation charts stored in the memory of the calculator that were drawn up previously with the solutions of the system of dynamic equations for the aircraft of this type, the general form of which is shown, for example, in [12, 13]. These charts are calculated for all possible dynamic states of the aircraft for various combinations of factors that affect the aircraft flight.

The linear approximation of the flown trajectory over a calculated time interval is performed using the method of moving average [14] by processing the horizontal coordinates and the aircraft height during this interval.

The use of the moving average method is required due to the need to reduce the effect of floating errors in the determination of the aircraft coordinates and height on the result obtained, as well as the effect of changes in the coordinates and height caused by atmospheric turbulence and random components of control signals.

After the notification of selecting only one runway is generated 16 and calculating the intended approach trajectory 9 based on the AND and the data on the preset distance from the runway threshold at the time of passing the height of 200 m above the runway for this type of aircraft, the space domain of permissible deviations from the intended trajectory 17 is determined (taking into account the aircraft dynamic capabilities and the accuracy performance of the onboard sensors), then the current deviations of the aircraft from the intended trajectory are calculated 18.

After the notification of selecting only one runway is generated, the predicted landing point is determined 19 as the point of intersection of the extended linearly approximated trajectory 12 with a horizontal plane passing through the middle point of the runway threshold, and then, the deviation of the said predicted landing point from the intended landing point is calculated 20.

Also, after the notification of selecting only one runway is generated, the parameters of the zone of permissible deviations of the predicted landing point from the intended one are calculated 21. This operation is performed according to the known onboard function, the dependence of the parameters of this zone from the distance to the runway threshold. The parameters of the said zone are defined as the sizes of the ellipse semi-axes with the center at the intended landing point lying in a horizontal plane passing through the middle point of the runway threshold. The area of the ellipse decreases as the distance from the aircraft to the runway threshold decreases. This function, on the basis of which the parameters of the zone of permissible deviations of the predicted landing point from the intended one are calculated, is determined in advance of statistical processing of the trajectories of successful approaches for aircraft of this type and is stored digitally in the onboard database.

If there is no confirmation of the runway selection, as well as in the case of failure to comply with the conditions of the said runway selection notification, an alert/warning is generated about unacceptable deviations of the aircraft from the intended trajectory and of the predicted landing point from the intended one 22. With this method, the aircraft descent to the boundary height above the runway is monitored 5. The described operations are continuously repeated until the boundary height is reached, beginning with the determination of the nearest runway 2 after receiving information from the onboard sensors 1. If the conditions for notification of selecting one runway are still met after the notification has been issued once, the said notification is no longer generated. If the above conditions are terminated, an alert/warning is generated about unacceptable deviations of the aircraft from the intended trajectory or about unacceptable deviations of the predicted landing point from the intended one (or both). If, then, as a result of the crew's actions, the conditions for notification of selecting a runway are met again, then the notification of the runway selection is generated again. Thus, the crew receives either a confirmation of the selection made earlier (in case the appearing unacceptable deviations were caused by control errors) or a notification of the selection of another runway (in case the appeared unacceptable deviations were caused by the maneuvering intentionally performed by the crew and conditioned by the contents of the notification of the wrong selection of the runway received earlier).

After reaching the boundary height, both the mentioned generation of the runway selection notification and the said generation of alerts/warnings of deviations from the required trajectory and the intended landing point are terminated.

The crew taking a decision on the continuation of the approach or on the missed approach takes into account the presence of generated notifications and alerts/warnings or their absence.

The use of the runway protection space refined by using the data on the allowable slopes of the intended approach trajectories for this type of aircraft to check the conditions for runway selection notifications makes it possible to increase the reliability of the said notification and thereby help the crew perform the landing better and safer.

Concerning the generation of alerts/warnings of deviations from the intended trajectory and the intended landing point, the following should be noted.

1. The selection of only one runway made before refining the protection space makes it possible to notify the crew of unacceptable deviations from the intended trajectory and the intended landing point, and thus, in a timely manner, to help the crew eliminate these deviations and expedite the notification of the runway selection for the approach.
2. Alerts/warnings of unacceptable deviations of the predicted landing point from the intended landing point should be formed only if the following conditions are met simultaneously:

   - The predicted landing point exists; that is, the predicted trajectory crosses a horizontal plane passing through the middle point of the runway threshold.
   - The nature of the unacceptable deviation is stable, i.e., the predicted landing point slightly changes its location in the displayed image (see Fig. 5.7) for a certain time (for example, for 10 s it moves on an area not exceeding 10% of the total area of the image or moves along the border of the picture by no more than 20% of its linear size), since a short-term "curvature" of the trajectory that can affect the rapid change in the result of linearization may be due to wind shear, step-wise errors of the position sensors, active control aimed at maintaining the aircraft on the intended approach trajectory, and other short-term factors.

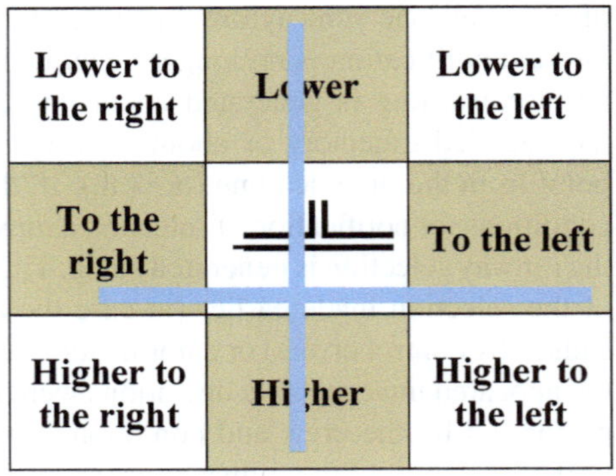

Fig. 5.6 Generation of alerts/warnings of unacceptable deviations of the aircraft from the intended trajectory

3. It is most effective to complement the alert/warning with a visual image. To generate alerts/warnings of unacceptable deviations of the aircraft from the intended trajectory, it is proposed to use the indication shown in Fig. 5.6. To generate alerts/warnings of unacceptable deviations of the predicted landing point from the intended one, it is proposed to use the indication shown in Fig. 5.7.

Following Fig. 5.6, information on the aircraft position can be obtained: above or below the intended trajectory; and on the deviation of the aircraft from the intended trajectory: to the right or left side (taking into account the nature of the vertical

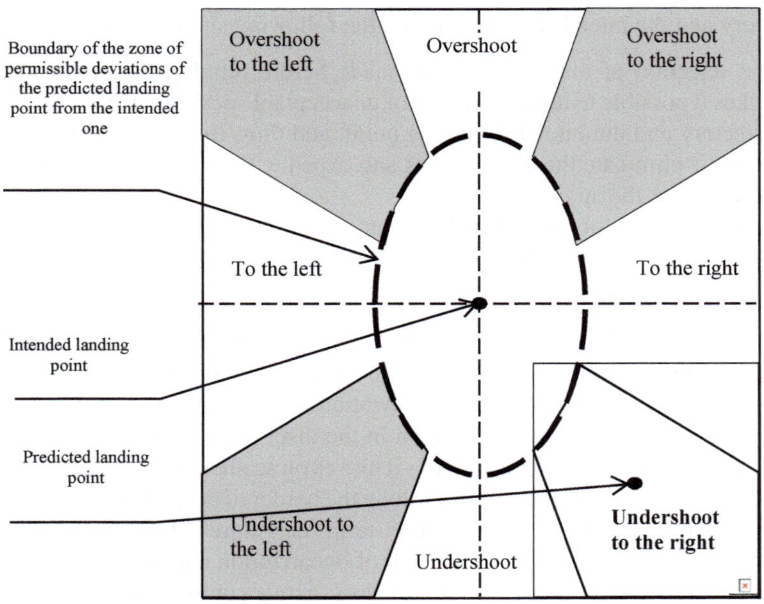

Fig. 5.7 Information on unacceptable deviations of the predicted landing point

deviation of the aircraft from the intended trajectory). The aircraft position is shown as a fixed symbol in the center of the screen. The displacement of the aircraft above or below the intended trajectory is shown by the position of the horizontal "bar," and the displacement of the aircraft to the right or left of the intended trajectory is shown by the position of the vertical "bar." In this case, the crossing point of the bars (the point of their intersection) shows the position of the intended trajectory.

The captions placed in eight areas of the screen show the content of alerts/warnings generated, if, when the conditions for their generation are met, the crossing of the bars (corresponding to the crossing of the traditional director used by the crew for performing instrumental landings) is within one of these areas.

The type of indication shown in Fig. 5.7 provides information on unacceptable deviations of the predicted landing point from the intended one: "overshot," "undershoot," "to the right," "to the left," "overshoot to the right," etc. The dimensions of the displayed ellipse that limits the range of permissible deviations of the predicted landing point remain unchanged with changes in the distance to the runway threshold, i.e., as the aircraft approaches the runway threshold, the displayed surface area decreases. If the deviations of the predicted landing point from the intended one are beyond the displayed area, the predicted landing point is displayed near the boundary in a place corresponding to its angular displacement with respect to the intended landing point. If the predicted landing point does not exist (the predicted trajectory does not intersect the horizontal plane passing through the middle point of the runway threshold), the predicted landing point is not displayed and an alert/warning of the deviation from the intended landing point is not generated in this case.

Figure 5.8 shows the boundary of the zone of permissible deviations from the intended landing point on the runway (solid line) [15] and possible boundaries of the zone of permissible deviations of the predicted landing point from the intended one (dashed lines) in the form of concentric ellipses with their centers at the intended landing point for two various values of the distance between the aircraft and the runway threshold.

Thus, the proposed method makes it possible to increase the reliability of aircraft landings by:

- generation of the notification of the runway selection for the approach adapted to the aircraft type;
- generation of an alert/warning of deviations of the aircraft position from the intended trajectory and deviations of the predicted landing point from the intended one when the conditions for generating a notification of the runway selection for the approach are not met (for non-instrument approaches);
- generation of a subsequent notification of the runway selection, if after the first notification the conditions for its generation were violated and then restored (for example, as a result of the crew's actions aimed at eliminating unacceptable deviations from the intended approach trajectory or eliminating unacceptable deviations of the predicted landing point from the intended one).

Let us perform an expert evaluation of the effect of the method considered on flight safety. This method can be most reasonably characterized using the factors

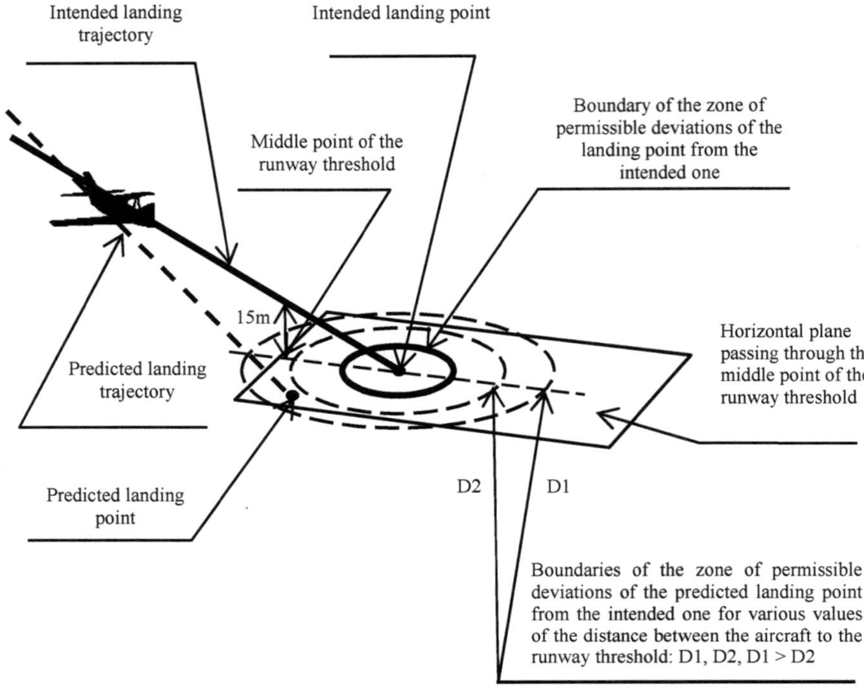

Fig. 5.8 Boundary of the zone of permissible deviations from the intended landing point and boundaries of the zone of permissible deviations of the predicted landing point on the runway

of ergonomics (F_{Erg}), reduction in the accident rate (F_{Acc}), and integration (F_{Int}) defined in Sect. 1.6. To assess the F_{Erg} value, we will determine the time that the pilot will spend on making a decision to land with and without using the proposed method.

It is known that alarms facilitate faster decision-making [16]. When performing an approach, the average time of recognition of the situation by the light panels is 13.5 s, and if the information is provided in writing or by voice, then this time is reduced to 0.5–5.5 s [16].

Then, it can be said that the implementation of the proposed method, regarding the generation of alarms for approaches on an unauthorized runway, allows saving from 8 to 13 s yielding the minimum F_{Erg} value = 13.5/5.5 = 2.4. It should be noted that the obtained $\mathbf{F_{Erg}}$ value can be expertly increased at least twice, taking into account the fact that in the considered method, the information about the aircraft performance is used as correctly as possible, including the determination of the permissible glide path angle, and also, unlike other known methods, in the considered method notifications are generated considering the current situation and can be refuted according to the actual situation. Thus, we can assume that $\mathbf{F_{Erg}}$ is about 4.8.

The accident rate reduction F_{Acc} can be estimated as follows. When performing an approach using the proposed method, the pilot receives complete information about the aircraft position relative to the glide path and the intended heading on the basis of the synthesized image, essentially corresponding to the type of displays of a conventional instrument landing system (see Fig. 5.6), as well as additional information on the deviation the predicted landing point from the intended one (see Fig. 5.7).

This information allows the pilot of monitoring the aircraft condition not only according to the indications of the navigation sensors, but, actually, also using the analog of the guidance system (instrument landing system). Given that the time for recognition of an emergency situation by the displays is 2–3 s, and the visual analysis of the external environment and navigation sensors requires an average of 10–20 s [16], we can expertly estimate the minimum value: $F_{\text{Acc}} = 20/3 \approx 6.7$.

Let us now estimate the integration factor $\mathbf{F_{Int}}$ that, in accordance with the definition (see Sect. 1.6), shows the relative increase in the probability of a safe flight before and after application of the described method. The characteristic features of the method considered above with the expert $\mathbf{F_{Int}}$ estimate makes it possible to consider it as an implementation of a virtual guidance system with the properties of an instrument landing system. When the SLS is used onboard, the implementation of similar functions in the CAS ensures duplication of the SLS.

In case of SLS failures, the CAS, in which the proposed method is implemented, can perform the functions of a backup (redundant) guidance (landing) system. Thus, the increase in the probability of performing a safe flight with the combined use of these two systems can be estimated expertly by the factor $\mathbf{F_{Int}} \approx 2$.

Let us now consider one more method that allows expanding the scope of integrated systems based on the SLS and CAS.

5.3 Method for Notifying of the Aircraft or UAV Position During the Landing and Roll-on Operation

The proposed method is a further development of the method described above to improve the landing safety and can be used after the pilot makes a decision about landing, touchdown, and during runway operations. The proposed method also provides determination and displaying of the aircraft deviations from the intended landing path (hereinafter—the intended glide path) and from the runway centerline during roll-on operations.

It should be noted that the issue of reducing the accuracy of determining the aircraft deviations from the intended glide path in angular units has not been considered earlier, given the known accuracy of the sensors for determining the aircraft coordinates. This problem is of significant importance for the SLS. In these systems, the error in calculating the angular deviations of the aircraft from the intended glide path traditionally shown on displays (for example, on the FNI) using the vertical

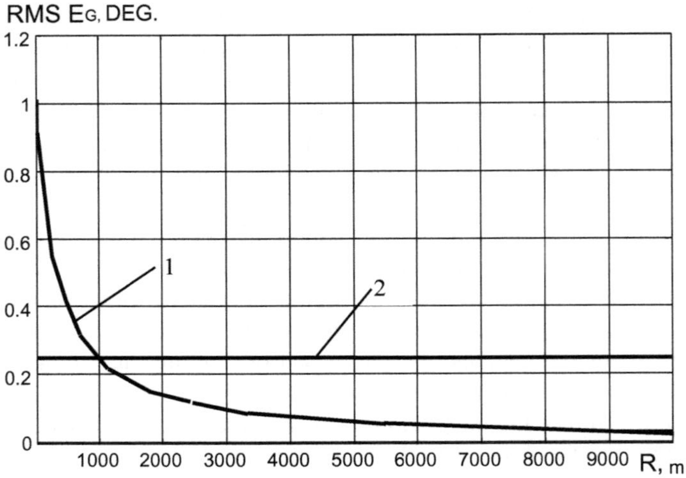

Fig. 5.9 Root-mean-square error in determining angular deviations from the glide path plane in the SLS

(heading) and horizontal (glide path) bars increases as the aircraft approaches the runway threshold. The latter is due to the fact that when using the GNSS navigation information, linear errors in determining the aircraft coordinates are practically constant during the approach.

Figure 5.9 shows the dependence (curve 1) of the root-mean-square error (RMS) of the angular deviation of the aircraft from the glide path plane with a pseudorange error $\sigma_\rho = 1.85$ m from the distance to the runway threshold (*R*), characteristic for the least favorable GNSS geometric situation for the SLS operation [17]. The line (2) in Fig. 5.9 corresponds to the requirements for Category I SLSs [18]. From Fig. 5.9, it can be seen that the RMS significantly increases when the aircraft approaches the runway and the use of angular deviations becomes impossible. At the same time, linear errors remain constant, which makes it possible to use them at short distances and even on the runway itself.

The idea of the proposed method consists in notifying the pilot of the aircraft position relative to the runway during the approach and during the movements after the touchdown. To generate a notification, onboard sensors provide information on the parameters of the aircraft dynamic state. Based on this information and the aeronautical database, an approach runway is selected. Further, the aircraft descent to the boundary height above each selected runway is monitored, and for runways above which the boundary height is not reached when the aircraft sinks, the conditions for the runway selection notification are monitored.

If these conditions are met, a corresponding notification is generated, and after selecting one runway and reaching the boundary height, the conditions of a notification of the distance to the runway end are monitored.

When the notification conditions are met, after passing half the runway, a notification is generated about the remaining distance to the runway end.

In this case, unlike the method described in Sect. 5.2, the presence of the "weight-on-wheels" signal is monitored and if it is not present, on the basis of the information on the aircraft coordinates, the information on the number of the runway designated and the information from their aeronautical database on the parameters of this runway, an intended glide path to the designated runway is calculated. During the landing process, the non-exceedance of the permissible range of the relative error in calculations of the aircraft deviation from the glide path plane in angular units is monitored by determining the boundary distance to the runway threshold, with the exceedance of which the relative error range exceeds the permissible value. Until the boundary distance is reached, aircraft deviations from the heading planes and the glide path are determined in angular units, and after reaching the boundary distance, the aforementioned deviations are determined in linear units. Further, in the presence of the "weight-on-wheels" signal and taking into account the information from the aeronautical database, the parameters and the aircraft dynamic capabilities, a zone of permissible deviations from the runway centerline is determined in linear units, lateral aircraft deviations from the runway centerline are calculated in linear units and a notification of the aforementioned deviations from the intended glide path and from the runway centerline is generated in the form of images on the displays with identical graphical landmarks (references).

This method of improving flight safety also proposes displaying the runway type corresponding to the actual type of runway as seen from the cockpit, below the boundary height and until stopping after rolling along the runway, in addition to displaying aircraft deviations from the intended glide path and the runway centerline; and if there is a zero lateral deviation the runway image has the form of an isosceles trapezoid with the axis of symmetry being a reflection of the runway centerline, and the middle point of its base lying on the same vertical line with the image of the bar with a zero lateral deviation, and if there is a nonzero lateral deviation, the runway image has the form of an non-isosceles trapezoid with the middle point of its base lying on the same vertical line with the image of the heading bar with a nonzero lateral deviation.

Following the "weight-on-wheels" signal and to facilitate control of the aircraft during the roll-on operation, it is proposed to determine and display a component of the actual acceleration vector parallel to the runway centerline and a component of the required acceleration vector parallel to the runway centerline, the value of which is determined from the calculation of the aircraft stop within the runway length. It is also proposed, in the presence of the actual acceleration corresponding to the aircraft braking during the roll-on operation, to continuously determine and display the projection of the calculated aircraft stop point on the runway centerline in such a way that the projection of the calculated stop point was shown in different colors depending on whether it is outside or within the runway length.

The component of the actual acceleration vector directed opposite to the landing heading of the runway and corresponding to the braking of the aircraft during the roll-on operation is proposed to be displayed in such a way that the end of this component when displayed is on the same horizontal level with the projection of the calculated point for the aircraft stop on the runway centerline. In this case, the component of the actual braking vector should be displayed with a solid arrow, and the component of the required braking vector—with a contour arrow.

When decelerating the aircraft, the displayed component of the actual acceleration vector and projection of the calculated stop point are proposed to be colored with the same color that changes when the calculated point exceeds the runway length.

In an emergency situation, the component of the actual acceleration vector coinciding with the landing heading of the runway in direction and corresponding to the acceleration of the aircraft during the roll-on operation, as well as the zero component displayed by a point and corresponding to the steady movement of the aircraft during the roll-on operation, is proposed to be colored with the same color as in the case the calculated stop point is beyond the runway length during braking.

It is proposed that, after passing through the middle point of the runway, the distance to the runway end be displayed at a specified interval.

During the roll-on operation, it is proposed to generate an alert/warning of exceeding the permissible value of the lateral deviation from the runway centerline in linear units, for example, in the form of a voice message.

The method under consideration is described in detail in [19]. The sequence of operations of the proposed method is shown in Fig. 5.10a, b and includes the following main operations:

- reception of information about the parameters of the aircraft dynamic state from its onboard sensors 1;
- selection of the runway based on the aeronautical database 2;
- control of aircraft descent to the boundary height 3;
- When the aircraft is above the boundary height, the notification conditions are monitored and, if they are met, a notification of the runway selection is generated 4; in addition to the above-mentioned boundary height, a zone of inhibition for the generation of the said alert/warning of the height is introduced that is defined by the type of aircraft and the need for other systems to generate a message saying that the aircraft has reached a certain height above the runway (for example, a collision avoidance system).
- If the runway selection is confirmed, the described operations are repeated until the aircraft sinks to the boundary height above each selected runway, below which no notification of the runway selection is generated.
- When the aircraft is below the boundary height, the conditions for notifying of the remaining distance to the runway end are monitored 5 and, when they are met, a notification is generated about the remaining distance, according to an a priori known list of values 6.

In the proposed method, new operations are introduced that are performed both above and below the boundary height. In Fig. 5.10, new operations are shown by thick solid lines, and the operations known from [19] are shown by thick dashed lines. It should be noted that known operations are implemented in [19] only when the

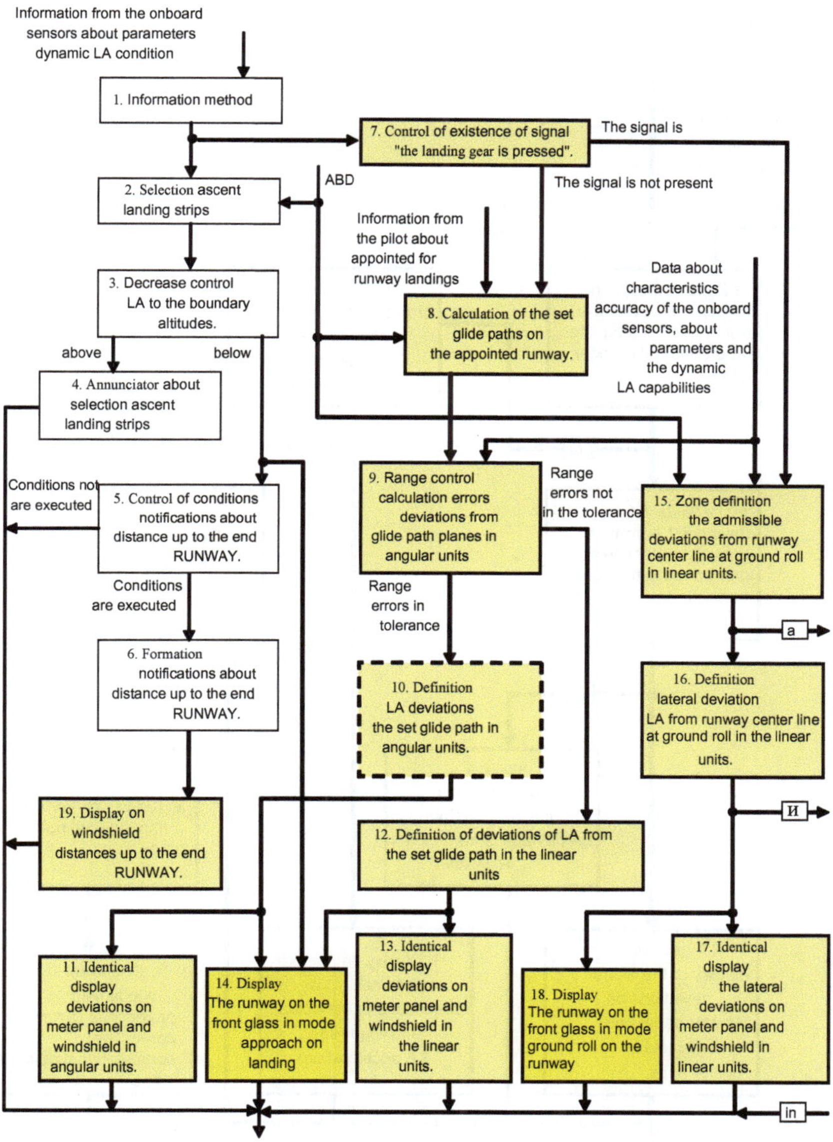

Fig. 5.10 **a** Sequence of operations for the proposed method, **b** sequence of operations for the proposed method (continued)

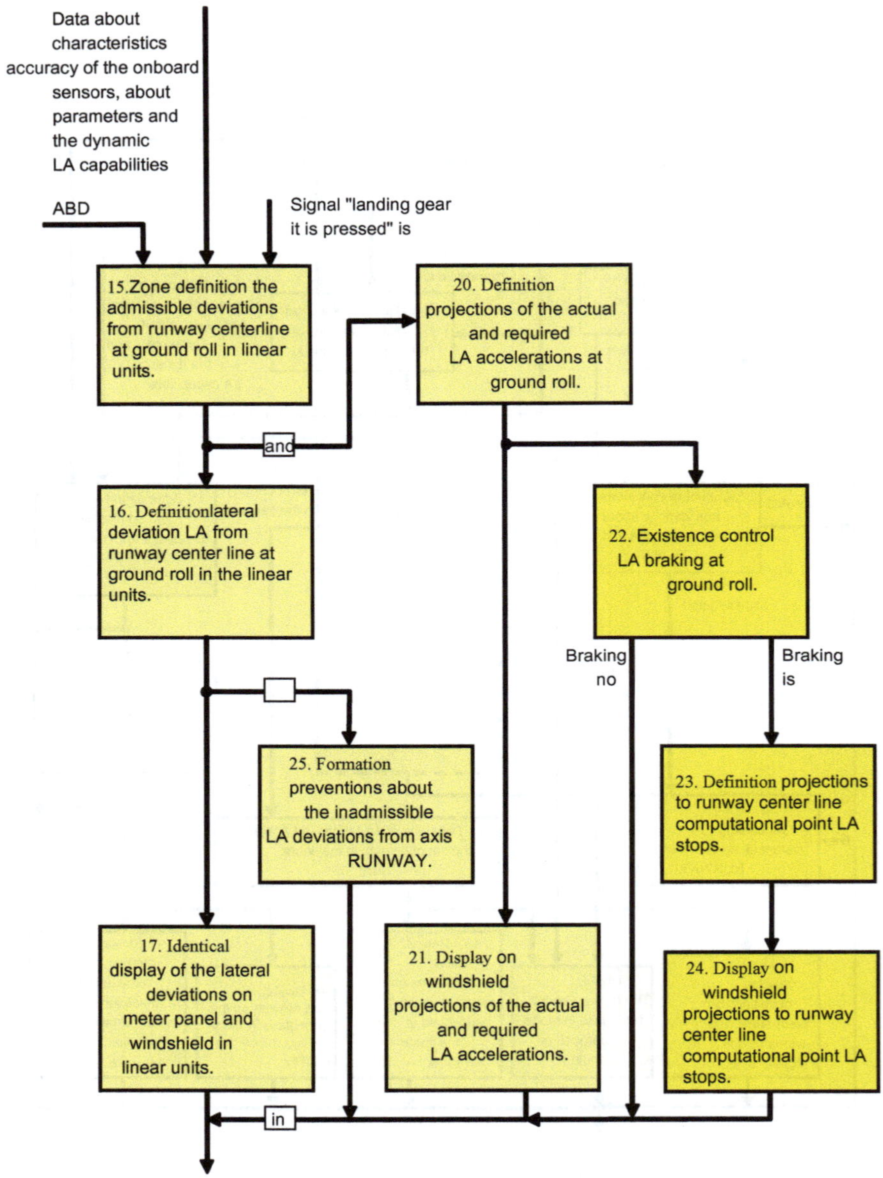

Fig. 5.10 (continued)

aircraft is above the boundary height, and in the method under consideration—when the aircraft is both above and below the boundary altitude.

Let us consider new operations for implementation of the new method (see Fig. 5.10). Initially, the presence of the "weight-on-wheels" signal (the aircraft is on the ground) is checked 7, and in the absence of this signal, an intended glide path is calculated on the designated runway 8.

The intended glide path lies in a vertical plane passing through the runway centerline. The intended glide path for the designated runway can be calculated in the same way as in the above method based on the aircraft height data at a certain distance from each selected runway.

Here, we consider a variant of calculating an intended landing glide path for a selected runway used in satellite-based landing systems (SLSs) and based on the use of the aeronautical database information on the runway designated for landing. The main advantage is the use for each runway of only the glide path that is recommended for it in the aeronavigation database, which ensures the maintenance of the stereotype in the crew's observation of the external references of the underlying surface when landing on a final segment of the glide path.

In the SLS, in order to maintain the continuity of displaying the landing signals generated by instrument systems (e.g., ILS), the landing signals are calculated in proportion to the angular deviations from the heading and glide path planes.

The deviations on the instrument panel, for example, on a horizontal situation indicator are displayed in such a way that the aircraft position is in the center of the screen.

In this case, negative values of the angular deviation of the aircraft from the heading plane (ε_{h}) and the angular deviation from the glide path plane (ε_{g}) are displayed by shifting the vertical (heading) and horizontal (glide path) bars to the left and upwards, respectively.

For the SLS, the stereometric view presented in Fig. 5.11 explains the method for calculating glide path parameters and angular deviations from the heading and glide path planes that are similar to the angular landing signals of ILS-like systems.

The glide path is established in the coordinate system *OXYZ* fixed relative to the Earth with the origin at the center of the Earth, the *X*-axis being directed to the point of intersection of the equator and the zero meridian, the *Z*-axis being directed along the Earth's rotational axis to the North Pole, and the *Y*-axis completes the system to the right system of vectors. The horizontal plane *FLS* is tangent to the earth ellipsoid at the reference point *L*.

The heading plane *LFK* perpendicular to the horizontal plane *FLS* passes through the runway centerline.

The glide path plane *BCM* slanting with respect to the horizontal plane passes through the glide path line and is perpendicular to the heading plane.

The dihedral angle between the glide path plane *BCM* and the horizontal plane *FLS* is equal to the glide path angle θ.

As was shown in [20, 21] and as can be seen from Fig. 5.11, the angular deviation of the aircraft from the heading plane can be determined from the formula:

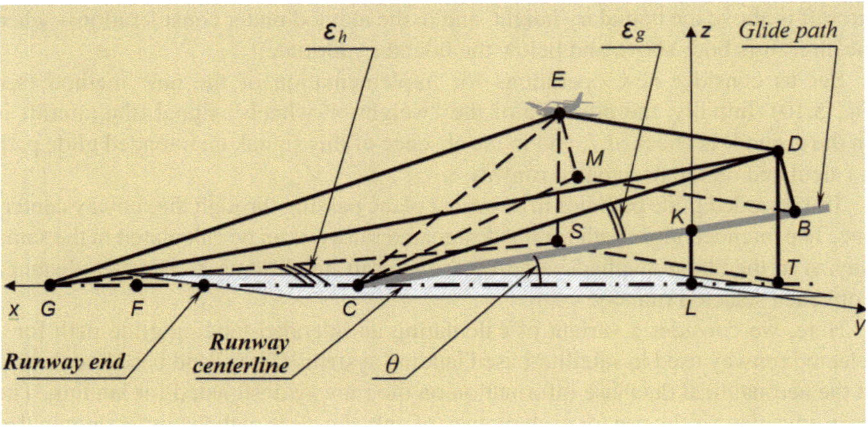

Designations: F is a point determining the direction of the runway centerline; G is a reference point for angular deviations from the heading plane; θ is the glide path angle; L is the middle point of the runway threshold; K is a reference point determining the position of the glide path; KL is a reference height determining the position of the glide path; C is the intersection point of the glide path and a horizontal plane passing through the middle point of the runway threshold; E is the point at which the aircraft is located; ε_h is an angular deviation of the aircraft from the heading plane; ε_g is an angular deviation of the aircraft from the glide path plane.

Fig. 5.11 To the calculation of the glide path and deviations from it in angular and linear units

$$g_{\mathrm{h}} = \arcsin \frac{ED}{SG}, \tag{5.1}$$

where *ED* is the distance from the point *E* (*X*, *Y*, *Z*) in which the aircraft is located to the heading plane, *SG* is a leg of the right triangle *ESG* with the right angle *S*.

The negative value of the angular deviation from the heading plane corresponds to the deviation of the aircraft to the right along the movement direction, in accordance with the axes of the local coordinate system *Lxyz* shown in Fig. 5.11, the origin of which is at the point *L*, the direction of the *x*-axis coincides with the landing course of the runway, the *z*-axis is directed vertically upwards, and the *y*-axis completes the system to the right-sided one.

The angular deviation of the aircraft from the glide path plane is determined by Formula [21]:

$$\varepsilon_{\mathrm{g}} = \arcsin \frac{EM}{CD}, \tag{5.2}$$

where *EM* is the distance from the aircraft position point to the glide path plane, and *CD* is a leg of the right triangle with the right angle *D*.

The negative value of the angular deviation from the glide path plane corresponds to the aircraft position below the glide path plane.

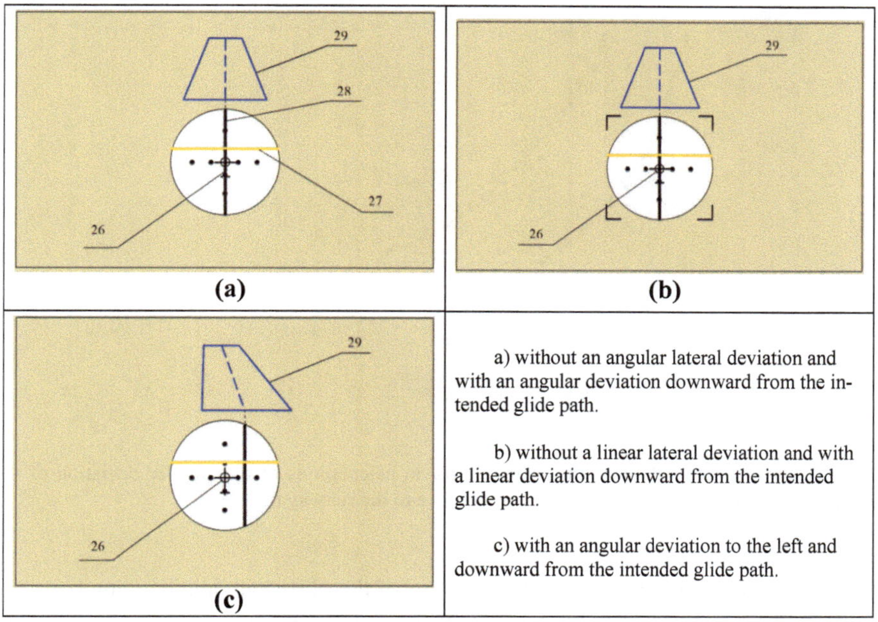

Fig. 5.12 Type of indication when the aircraft is in the final approach segment

After calculating the intended glide path, the non-exceedance of the permissible range of the relative error in calculations of deviations from the intended glide path in angular units (taking into account the data on the accuracy of onboard sensors) is monitored (see 9 in Fig. 5.10). If there is an error in this range, the current aircraft deviations from the intended glide path in angular units are determined, as in [19] (see 10 in Fig. 5.10), and these deviations are displayed in angular units (see 11 in Fig. 5.10). If there is an error outside this range, the aircraft deviations from the intended glide path are determined in linear units (see 12 in Fig. 5.10) and displayed also in linear units (see 13 in Fig. 5.10). In both cases, below the boundary height, the runway is displayed in the approach mode (see 14 in Fig. 5.10).

The type of displaying on the final approach segment using angular or linear deviations is shown in Fig. 5.12.

Let us analyze the conditions of switching from angular deviations to linear deviations from the intended glide path.

The magnitude of the angular deviations of the aircraft from the heading and glide path planes for the same linear deviations increases as the aircraft approaches the reference points of these angular deviations. For the SLS, these points are usually located in places corresponding to the locations of the localizer and glide slope beacons of ILS-like instrument landing systems.

Therefore, the presence of constant root-mean-square errors (RMS) in determining horizontal coordinates and height by the navigation system of the aircraft will

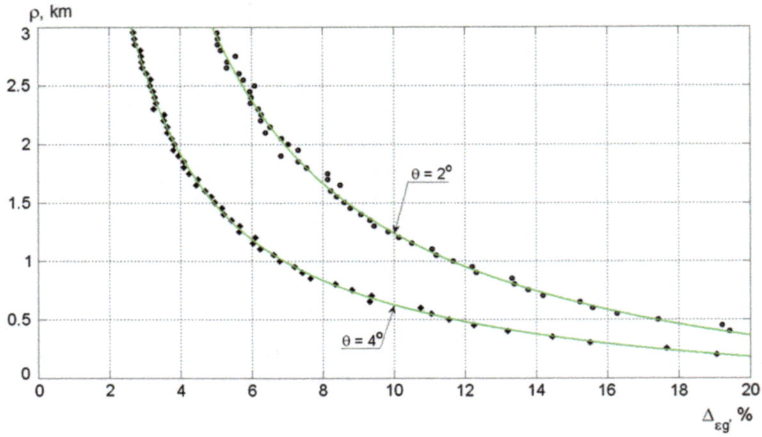

Fig. 5.13 Allowable values of the relative error in calculations of the angular deviation of the aircraft from the glide path plane from the distance to the runway threshold

lead to an increase in the RMS of deviations of the glide path and heading bars, the intersection point of which reflects the position of the intended glide path relative to the current aircraft position.

The switching from angular deviations from the heading and glide path planes to linear deviations from these planes (hereinafter referred to as the switching to the linear deviation display mode) should be made simultaneously for convenience of the pilot during manual control or when these deviations are sent to the aircraft ACS.

To determine the distance at which it is advisable to switch to linear deviations from the heading and glide path planes, taking into account the permissible range of relative errors in calculations of the angular deviation of the aircraft from the glide path plane (hereinafter referred to as PRR) depending on GNSS errors, statistical modeling was performed. The use of statistical modeling was due to the fact that the analytical expression for the PRR is extremely difficult to analyze.

The modeling was conducted assuming constant measurement errors in the GNSS over the simulated time interval. In view of the fact that the reference point for the angular deviations from the heading plane is farther from the runway threshold than the reference point for the angular deviations from the glide path plane, the distance at which it is reasonable to switch to the linear deviation display mode is determined based on the results of the PRR behavior study.

Figure 5.13 shows the statistical dependencies of the PRR value ($\Delta_{\varepsilon g}$) on the distance to the runway threshold (ρ) obtained from the simulation:

$$\Delta_{\varepsilon g} = f(\rho), \tag{5.3}$$

Table 5.1 Factors of the approximating function

	a	b	c	d
$\theta = 2^{o}$	3.708	−0.116	9.845	−0.487
$\theta = 4^{o}$	7.279	−0.598	2.035	−0.121

The PRR value was determined in accordance with expression:

$$\Delta_{\varepsilon g} = \frac{2\sigma_{\varepsilon g}}{\varepsilon_{g\_max}} \times 100\%, \tag{5.4}$$

where $\sigma_{\varepsilon g}$ is the RMS of the angular deviation of the aircraft from the glide path plane due to the error in determining the aircraft height, and ε_{g_max} is the half of the display range the angular deviations of the aircraft from the glide path plane used in the standard onboard flight navigation instruments (FNIs).

The quantity ε_{g_max} is defined as [18]

$$\varepsilon_{g\_max} = 0.25 \cdot \theta, \tag{5.5}$$

where θ is the glide path angle.

The results of the statistical modeling presented in Fig. 5.13 are obtained for the minimum (2°) and maximum (4°) values of the glide path angle θ usually used in ILS-like instrument landing systems. It was assumed in the modeling that the law of the coordinate measurement error distribution is normal and characterized by the error in measuring horizontal (0.5 m) and vertical (0.75 m) coordinates (RMS), which corresponds to the differential mode of the GNSS operation [12].

For convenience of use in real onboard devices, the results of the statistical modeling presented in Fig. 5.13 were approximated using the least squares method [14] by an analytic function of the form:

$$\rho = a \cdot e^{b \cdot \Delta_{\varepsilon g}} + c \cdot e^{d \cdot \Delta_{\varepsilon g}}, \tag{5.6}$$

where ρ is the distance from the aircraft to the landing threshold point of the runway and $\Delta_{\varepsilon g}$ is the PRR value; *a*, *b*, *c*, and *d* are factors, the values of which are given in Table 5.1.

The empirical Formula (5.5) makes it possible to determine the distance at which it is reasonable to switch to displaying linear deviations from the heading and glide path planes, depending on the permissible $\Delta_{\varepsilon g}$ value, taking into account the RMS coordinates determined by the navigation system of the aircraft and the glide path angle that determines the range of the scale used to display angular deviations of the aircraft from the glide path plane (see 5.4).

With $\Delta_{\varepsilon g} = 10\%$ in the differential mode of GNSS measurements, the distance at which it is advisable to switch to the linear deviation display mode lies in the range from 625 to 1238 m, depending on the glide path angle, the value of which lies in the range from 4° to 2°, respectively (see Fig. 5.7).

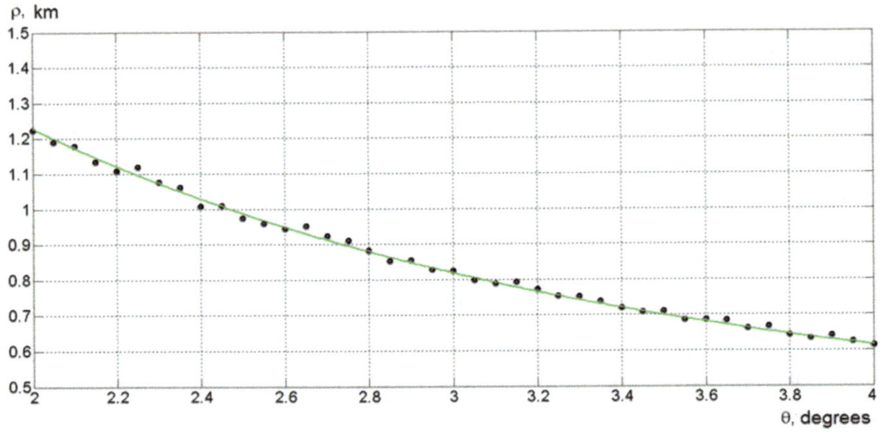

Fig. 5.14 Dependence of the distance from the aircraft to the runway threshold on the glide path angle for a PRR value equal to 10%

Table 5.2 Factors of the approximating function

	a	b	c	d
$\Delta_{\varepsilon g} = 10\%$, $\sigma_H = 0.75$ (m)	3.306	−0.767	0.574	−0.055

Dependence of the distance from the aircraft to the runway threshold on the glide path angle for a constant PRR value equal to 10%, with the permissible RMS value determining the coordinates in the GNSS is shown in Fig. 5.14.

This dependence makes it possible to determine the distance at which it is reasonable to switch to displaying linear deviations from the heading and glide path planes, taking into account the RMS coordinates determined by the navigation system of the aircraft and depending on the glide path angle, on which, in turn, the range of angular deviation displaying depends. To approximate the dependence shown in Fig. 5.8, the following a function was used:

$$\rho = a \cdot e^{b \cdot \theta} + c \cdot e^{d \cdot \theta}, \tag{5.7}$$

where ρ is the distance from the aircraft to the runway threshold; θ is the glide path angle on which the range of angular deviation displaying depends; a, b, c, and d are factors obtained using the least squares method. The calculated values of the factors are given in Table 5.2.

After switching to the linear deviation display mode, it is proposed to calculate the deviation from the heading plane as the distance *ED* from the aircraft to the heading plane and the deviation from the glide path plane as the distance *EM* from the aircraft to the glide path plane (Fig. 5.11).

Table 5.3 Ranges for displaying linear deviations

	$\theta = 2°$ ($\rho =$ 1238 m)	$\theta = 3°$ ($\rho =$ 826 m)	$\theta = 4°$ ($\rho =$ 625 m)
z_{max} (m)	15	15	15

Signs of linear deviations correspond to signs of angular deviations, namely when the aircraft is below the glide path plane and to the right from the heading plane (in the direction of movement), the linear deviations are negative.

At the switching to the linear deviation display mode, the scale changes in such a way that the glide path bar, the deviation of which from the center of the screen is proportional to the angular deviation, remains at the same place on the screen. This is achieved by recalculating the scale of displaying deviations from the glide path plane by following formula:

$$Z_{gmax} = BC_{max} \cdot \mathrm{tg}\left|\varepsilon_{gmax}\right|, \tag{5.8}$$

where Z_{gmax} is half of the range of displaying linear deviations of the aircraft from the glide path plane, and BC_{max} is the distance from the intersection point C of the glide path with the runway centerline to point B on the glide path (the Z_{gmax} value will be the maximum one when the aircraft is on the glide path at the boundary distance ρ to the runway threshold calculated by Formula (5.7), at the moment of switching to the linear deviation display mode); ε_{gmax} is half the range of displaying angular deviations of the aircraft from the glide path plane in degrees.

The use of the Z_{gmax} value, depending on the glide path angle θ as a half of the FNI scale for displaying the linear deviation of the aircraft from the glide path plane, results in a dependence of the display scale on the θ value. Table 5.3 shows various values of the ranges for displaying the height linear deviation for various values of possible glide path angles.

As can be seen from Table 5.3, the scales of displaying the height linear deviation of the aircraft from the intended glide path for various values of possible glide path angles are approximately the same; therefore, it is advisable to display linear deviations from the glide path with an angle of 2° to 4° typical for most aircraft, in a single scale ±15 m.

The half of the range of displaying linear deviations of the aircraft from the heading plane is calculated from the condition of the heading course bar immobility at the moment of switching to the linear deviation display mode by the formula:

$$y_{max} = GT_{max}\,\mathrm{tg}|\varepsilon_{h\,max}|, \tag{5.9}$$

where GT_{max} is the distance from reference point G of the angular deviation from the heading plane to point T (the distance GT will be maximum with $GT = GS = GT_{max}$, when the aircraft is on the glide path at the boundary distance ρ to the runway threshold calculated by the Formula (5.6), at the moment of switching to the linear

Table 5.4 Ratios of scales of linear and lateral deviations

$y_{\max}$ (m)	$\theta = 2$ ($\rho = 1238$ m)	$\theta = 3$ ($\rho = 826.1$ m)	$\theta = 4$ ($\rho = 625$ m)
$L_{\mathrm{RWY}} = 500$	330	255	218
$L_{\mathrm{RWY}} = 5405$	137	126	121

deviation display mode); $\varepsilon_{\mathrm{hmax}}$ is half the range of displaying angular deviations of the aircraft from the heading plane.

The half of the range of displaying angular deviations of the aircraft from the heading plane is determined by the formula

$$\varepsilon_{\mathrm{h\_max}} = \operatorname{arctg}\left(\frac{C_w}{LG}\right), \tag{5.10}$$

where C_w is the width of the course; LG is calculated as the distance from the runway threshold point $L\,(X_L, Y_L, Z_L)$ to the reference point of heading angles $G\,(X_G, Y_G, Z_G)$ that is similar to the location of the localiser beacon in the ILS system.

Thus,

$$y_{\max} = GT_{\max}\mathrm{tg}|\varepsilon_{\mathrm{h}}{\max}| = \frac{GT_{\max} \cdot C_w}{LG}. \tag{5.11}$$

The use of the $y_{\max}$ value that depends on the location of the reference point of the angular deviations from the heading plane G, and hence on the runway length, as a half of the scale for displaying linear deviations of the aircraft from the heading plane results in a dependence of the display scale on the LG value (9). Table 5.4 shows the ratios of the scales of the linear lateral deviation displaying ensuring the immobility of the heading bar at the moment of switching to the linear deviation display mode for the minimum and maximum runway lengths available in the aeronautical database.

It is shown in Table 5.4 that the scales of displaying the lateral linear deviation of the aircraft from the intended glide path for various values of the possible length of the LG segment vary significantly.

In view of the fact that it is convenient to use a single scale for displaying linear lateral deviations independent of the runway length, it is suggested to always use a single range of ±30 m corresponding to the maximum runway width for most aerodromes to display linear lateral deviations.

At the same time, at the moment of switching to the linear deviation display mode with the heading bar displaying the angular deviation from the heading plane, the position of the heading bar will in general change sharply, but the heading bar will remain on the same side with respect to the vertical symmetry axis of the screen passing through the center of the screen corresponding to the display of the aircraft position, that is, the sign of the lateral linear deviation of the aircraft from the heading plane will be the same as the sign of the angular deviation of the aircraft from this plane.

After the landing of the aircraft, i.e., in the presence of the "weight-on-wheels" signal, using the data on the aircraft dynamic capabilities and the accuracy performance of the onboard sensors, this method provides determination of the zone of permissible deviations from the runway centerline during the roll-on operation in linear units 15, determination of deviations 16, displaying the lateral deviation in linear units 17, and displaying the runway in the roll-on mode 18 (see Fig. 5.10).

The display also shows the distance to the runway end. The displayed distance values correspond to the list of values that are announced by voice in [22], and the duration of displaying the distance values corresponds to the duration of the voice announcement.

Such a measure is aimed at increasing the reliability of informing the crew about the distance to the runway end and may be the only source of the information mentioned if there are malfunctions in the onboard voice information reporting system.

After the zone of permissible deviations from the runway centerline (see 15 in Fig. 5.10) has been determined, the projections of the aircraft actual and required acceleration vectors during the roll-on operation onto the runway centerline are made 20 and the mentioned projections of the vectors are displayed 21 (see Fig. 5.15). Simultaneously with the determination of these vector projections, the direction and magnitude of the aircraft actual acceleration projection onto the runway centerline during the roll-on operation are analyzed 22 and, in the case of braking, the projection of the calculated aircraft stop point on the runway centerline is made 23, and this projection of the calculated aircraft stop point is displayed 24.

After determining the linear lateral deviation of the aircraft from the runway centerline, taking into account the previously determined zone of permissible deviations from the runway centerline during the roll-on operation, when the corresponding conditions are met for the ratio of the linear lateral deviation and its rate of change (see Fig. 5.16), an alert/warning of unacceptable deviations of the aircraft from the runway centerline is generated.

The method proposed above makes it possible to improve pilot's awareness of the aircraft position.

From the image in Fig. 5.12, it is possible to obtain information about the aircraft position above or below the intended glide path and the deviation of the aircraft from the intended glide path. The aircraft position is shown by a fixed symbol in the center of the screen 26. The vertical displacement of the aircraft above or below the intended glide path is shown by the position of the glide path bar 27, and the lateral displacement of the aircraft to the right or left of the intended glide path is shown by the position of the heading bar 28. Their crossing (intersection) point shows the position of the intended glide path. At the landing segment on the runway 29 below the boundary height and until the "weight-on-wheels" signal, the display of the aircraft deviations from the intended glide path is identical to display of the same deviations on the instrument panel, which simplifies the task of the alternate use of various indicators by the pilot.

Replacement, at a certain distance from the runway threshold, of the displayed deviations from the intended glide path in angular units by the same displayed deviations in linear units, as well as their determination using accurate measurements of

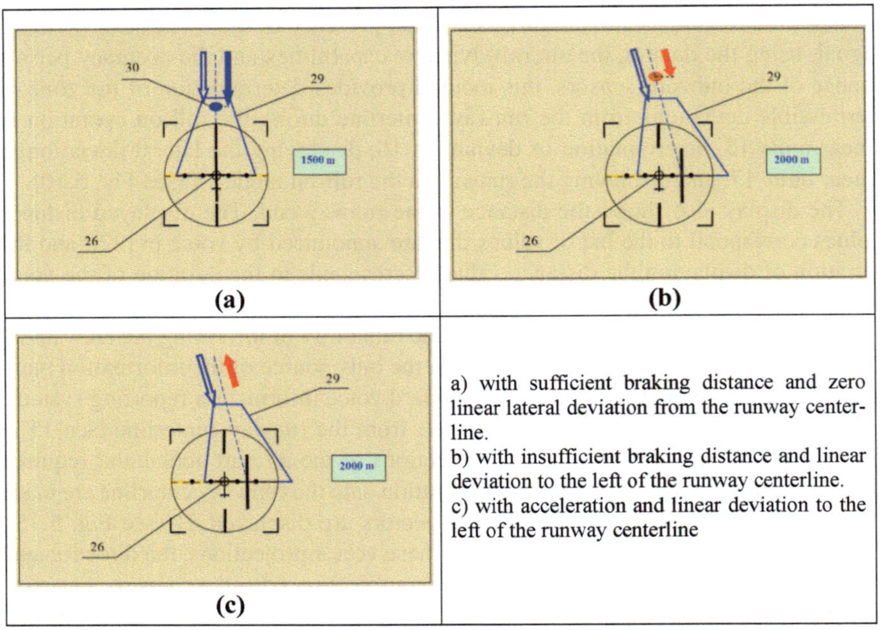

Fig. 5.15 Indication during the aircraft roll-on operation

the aircraft coordinates and the aeronautical database, makes it possible to extend the information utility of these deviations to control the aircraft until the end of its roll-on operation up to the calculated stop point.

The data displayed in accordance with Fig. 5.15 inform on the following:

- the value of the aircraft lateral deviation from the runway centerline during the roll-on operation after the landing gear compression;
- the ratio of the projected vector values of the required and actual braking of the aircraft along the runway centerline; the projection of the required braking vector is displayed with an uncolored arrow pointing downwards, the length of which is selected so that the horizontal line drawn through the end of the vector of said projection intersects the runway image on the head-up display near the displayed runway end, and the projection of the actual braking vector is displayed: (a) with a blue arrow pointing downwards if the actual braking along the runway centerline is not less than the required one; (b) with a red arrow pointing downwards, if the actual braking along the runway centerline is less than required one and there is a hazard of aircraft overrunning beyond the runway length; (c) with a red arrow pointing upwards, if the aircraft is not decelerated, but accelerates while moving along the runway (in case of an emergency situation onboard) and the calculated stop point is not displayed (see Fig. 5.15); (d) is not displayed at all, if the aircraft is moving along the runway at a constant speed (the calculated stop point is also not displayed in this case);

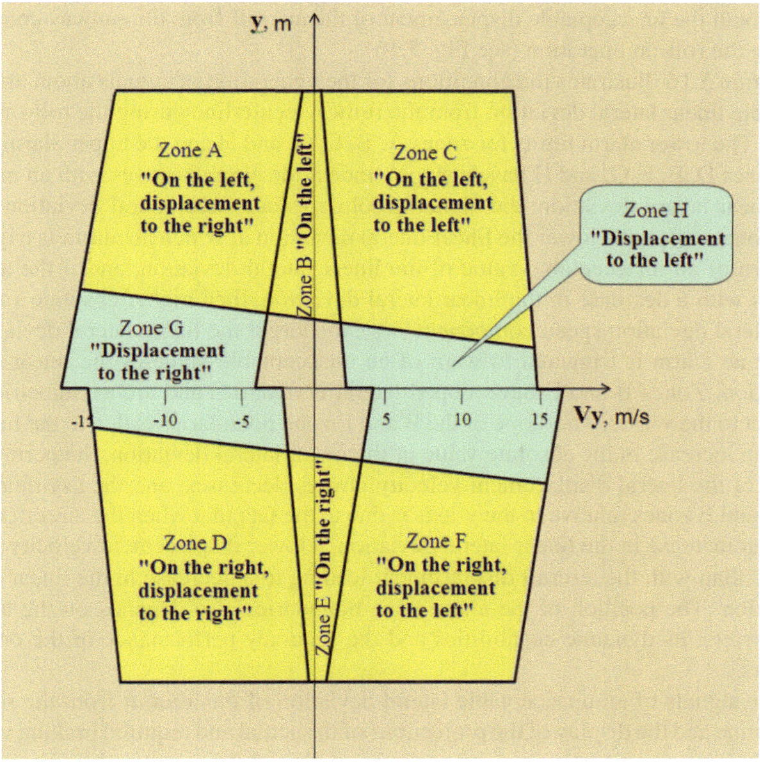

Fig. 5.16 Zones of alert/warning signal triggering when the value of the linear lateral deviation from the runway centerline during the roll-on operation in the coordinate system is unacceptable: the linear lateral deviation is the linear lateral deviation rate of change

- the projected position of the aircraft stop point on the runway centerline, while one of the usable colors of this point (for example, blue) shall indicate that there is no hazard of aircraft overrunning beyond the runway length, and another color (for example, red) indicates the presence of such a hazard.

Note that the projection of the calculated aircraft stop point onto the runway centerline, which is located outside the runway length, is always colored in color indicating the hazard of aircraft overrunning the runway, for example, in red (see Fig. 5.15b), and the projection of the calculated aircraft stop point located within the runway length is always shown in color indicating there is no hazard of overrunning the runway, for example, in blue (see Fig. 5.15a), even if there is a hazard of overrunning beyond the runway width. It is proposed that the pilot should assess the hazard of overrunning beyond the runway width by the position the aircraft heading bar relative to the runway centerline indicated during the roll-on operation along the runway on the head-up display and instrument panel and also by generated signals about the unacceptable displacement of the aircraft from the runway centerline during the roll-on operation (see Fig. 5.16).

Figure 5.16 illustrates the conditions for the generation of signals about an unacceptable linear lateral deviation from the runway centerline during the roll-on operation. The lower alarm limits for zones A, B, C, G, and H and the upper alarm limits for zones D, E, F, G, and H have a slope, since if the aircraft moves with an increasing linear lateral deviation, the higher absolute value of the lateral deviation speed component (V_y), the lower the linear lateral deviation at which an alarm is triggered to warn of an unacceptable value of the linear lateral deviation, and if the aircraft moves with a decrease in the linear lateral deviation, then higher absolute value of the lateral deviation speed component (V_y), the larger the linear lateral deviation at which an alarm is triggered to warn of an unacceptable value of the linear lateral deviation. Zones B and E have sloped lateral boundaries and are asymmetric with respect to the **y**-axis. The slope of the B and E zone boundaries is due to the fact that with an increase in the absolute value of the linear lateral deviation, the permissible value of the lateral displacement velocity always decreases, and the asymmetry of the B and E zones relative to the **y**-axis is due to the fact that when the aircraft moves with an increase in the linear lateral deviation, a lower displacement velocity is permitted than with the aircraft displacement leading to a decrease in the linear lateral deviation. The position of the alarm zone boundaries also depends on the aircraft parameters, its dynamic capabilities and the accuracy performance of the onboard sensors.

The signals of an unacceptable lateral deviation of the aircraft from the runway centerline and the display of the projections of the actual and required braking vectors onto the runway centerline, as well as the projection of the calculated stop point, will help the pilot with timely estimation of the aircraft movement parameters during the roll-on operation and, if necessary, with timely corrective adjustments to the aircraft control required to complete the aircraft roll-on operation within the runway.

Thus, the proposed method makes it possible to increase the safety of aircraft landing, including the roll-on operation with the following:

- timely switching from the display of angular deviations of the aircraft from the heading and glide path planes to the display of linear deviations from the mentioned planes;
- displays in identical geometric references of deviations of the aircraft from the intended glide path up to the moment of the landing gear compression and of deviations from the runway centerline during the roll-on operation;
- generation of an alert/warning about an unacceptable lateral deviation from the runway centerline during the roll-on operation;
- display of the announced (by voice) values of the distance remaining to the runway end;
- display of the projections of the required and actual braking vectors parallel to the runway centerline;
- display of the projection of the calculated aircraft stop point onto the runway centerline.

Let us evaluate the effect of the proposed method on flight safety, using the factors defined in Sect. 1.6 of this paper. As an initial level, we will consider an instrument SLS, which does not use linear deviations from the landing path and does not display deviations from the runway centerline during the roll-on operation.

To evaluate the method proposed above, we use the factors introduced in Sect. 1.6: $\mathbf{F}_{\mathbf{Acc}}$, $\mathbf{F}_{\mathbf{Cont}}$, $\mathbf{F}_{\mathbf{Erg}}$, and $\mathbf{F}_{\mathbf{Int}}$.

Consider the continuity factor $\mathbf{F}_{\mathbf{Cont}}$. As noted above, the switching to linear deviations provides the pilot with information on deviations from the landing path with an accuracy corresponding to the GNSS accuracy in the differential mode. The accuracy of determining angular deviations from the glide path at distances less than 1 km from the runway threshold may already be insufficient to meet the relevant requirements. The switching to the linear scale makes it possible to maintain the display of deviations with accuracy sufficient for landing and rolling-on along the runway. When moving at a speed of 75 m/s for the last 13 s from the flight to the landing and another 30 s of the rolling-on, the pilot will be able to obtain instrument information about the deviation from the intended trajectory. Thus, if we take 150 s for the average approach time, the use of the proposed method will increase the time of continuous reception of landing information from $150 - 30 - 13 = 107$ s to 150 s. Then, we obtain $\mathbf{F}_{\mathbf{Cont}} = 150/107 \approx 1.4$.

Consider the ergonomics factor $\mathbf{F}_{\mathbf{Erg}}$ and estimate the time that the pilot will spend evaluating the current situation using the proposed method. As is known [16], the time for switching attention is 3 s, the analysis of the external environment can take up to 10 s, and the decision on the nature of the necessary actions required from 5 to 20 s. The availability of instrument indication during the approach, landing and rolling-on with the method under consideration does not require switching attention and a long analysis of the external environment, and also provides the pilot with information about the nature of the movement along the runway, the real nature of the braking performance and the predicted stop point. Considering that the entire duration of the landing and roll-on phase is about 43 s (see above), we obtain $\mathbf{F}_{\mathbf{Erg}} = 43/(3 + 10 + 5) \approx 2.4$.

The integration factor $\mathbf{F}_{\mathbf{Int}}$ is estimated as follows. The method proposed above is most reasonable to use when both the SLS and CAS are available onboard the aircraft since the latter contains aeronautical databases with all the characteristics of the runway, as well as artificial obstacles in the terminal area, aircraft performance characteristics, etc., and the SLS has exact coordinates of the aircraft. Then, $\mathbf{F}_{\mathbf{Int}} = \mathbf{F}_{\mathbf{Cont}}\,\mathbf{F}_{\mathbf{Erg}} \approx 1.4 \cdot 2.4 = 3.4$.

Taking into account the fact that in the development of the proposed method, the zones warning of unacceptable value of the linear lateral deviation from the runway centerline during the roll-on operation were introduced for the first time, and also that the type of indication with the predicted stop point was proposed for the first time, the factor of relative decrease in the accident rate $\mathbf{F}_{\mathbf{Acc}}$ can be estimated by a large number that corresponds to the essential values of this factor obtained in this paper. It can be assumed that $\mathbf{F}_{\mathbf{Acc}} = 5$ for the case of an aircraft moving on the runway.

Thus, the above method for notifying of the aircraft position during the landing and roll-on phases makes it possible to significantly improve flight safety by providing the crew with additional information on the aircraft current state and predicted position. It should also be noted that these estimates do not take into account possible erroneous actions of the pilot in the absence of information, the generation of which is ensured by using the proposed method.

Let us now turn to the analysis of the flight safety improvement when using integrated systems, which, in essence, will be the systematization and generalization of the results in the previous three sections.

5.4 Assessment of Flight Safety and Efficiency Improvements with the Use of Integrated Systems

Let us perform a generalized quantitative expert evaluation of the factors defined in Sect. 1.6, which characterize the relative increase in flight safety with the use of an integrated system based on the SLS and CAS.

Let us consider the factor of relative increase in the accuracy of determining the navigational parameters $\mathbf{F_A}$. The use of the corrected coordinates in the SLS makes it possible to reduce the error in determining the deviation of the aircraft from the landing glide path to a value of the order of 1 m (with a probability of 95%). In ILS-like Category I instrument landing systems, this error is [18] no more than 10 m near the decision height point regarding execution of a landing or a missed approach. Then, taking into account the true SLS accuracy, we can estimate $\mathbf{F_A} \approx 10/1 = 10$. Since, in accordance with the general requirements for Category I SLSs, the requirements are below the real accuracy of determining the coordinates in the differential mode (see [18]), then, within the minimum requirements, it is possible to estimate the $\mathbf{F_A}$ value for the glide path channel by the value $\mathbf{F_A} \approx 4/1 = 4$, and for the heading channel by the value $\mathbf{F_A} \approx 16/1 = 16$.

The generation of warning alarms in the CAS mode of "excessive deviation from the glide path" begins with deviations of 0.05 DDM (which corresponds to ≈10 m near the decision height point). Even with an allowable linear error of 4 m in determining the deviation from the glide path in the SLS, it is possible to estimate the ergonomics factor $\mathbf{F_{Erg}}$ as the ratio of the false alarm probability with and without the use of SLS data in the CAS: $\mathbf{F_{Erg}} \approx 4/1 = 4$, i.e., the number of false alarms will be reduced by four times.

The use of corrected SLS data that can be used in the CAS for aircraft flying in the LAAS coverage area, improves flight safety by refining both the parameters used to generate alert and warning signals in the basic CAS modes, or when using early warning or notification modes.

For definiteness, we assume that the CAS uses SLS coordinates and velocities that are characterized by practically constant errors of 1 m and 0.03 m/s (with a probability of 0.95).

Then, if we use the SLS vertical speed data in CAS mode 1 (excessive descent rate) instead of the data from variometers having an error of the order of 0.5–1.0 m/s at low descent rates and 2–5 m/s at high descent rates [23, 24], we obtain the decrease in the false alarm rate proportional to the speed measurement error for approaches at altitudes below 100 m. The minimum value of the ergonomics factor will be $\mathbf{F}_{\mathbf{Erg}} \approx 0.5/0.03 = 17.7$.

If the CAS altitude with an error of 1 m is used in the CAS with (or instead) the barometric altitude determined with an error of the order of 10 m [23, 24] in mode 2 (excessive ground proximity), then the expert estimate gives the minimum value of the ergonomics factor $\mathbf{F}_{\mathbf{Erg}} \approx 10/1 = 10.0$.

For CAS mode 7 (early warning), it is possible to estimate the value of the accident rate reduction coefficient $\mathbf{F}_{\mathbf{Acc}}$ as a ratio of the volumes of the aircraft protection space generated from data of the GNSS operating in standard and differential modes. This ratio is proportional to the ratio of products of errors, respectively, of the horizontal coordinates, height and ground speed. Then, for approaches at a speed of W = 75 m/s, it is possible to estimate the change in the volume of the protection space extending over a time interval $\Delta t = 60$ s ahead of the aircraft and having the minimum allowable obstacle clearance height $H_{\mathrm{ma}} = 30$ m (see Sect. 1.4) as follows:

$$\begin{aligned} F_{\mathrm{Acc}} &= (((W+\Delta W_{\mathrm{s}}) \cdot \Delta t+\Delta X_{\mathrm{s}}) \cdot (H_{\mathrm{ma}} - \Delta H_{\mathrm{s}}))/(((W+\Delta W_{\mathrm{d}}) \cdot \Delta t+\Delta X_{\mathrm{d}}) \cdot (H_{\mathrm{ma}} - \Delta H_{\mathrm{d}})) \\ &= (((75 + 0.3) \cdot 60 + 13) \cdot (30 + 22))/(((75 + 0.03) \cdot 60 + 1) \cdot (30 + 1)) \approx 1.7, \end{aligned}$$

where ΔW_{s}, ΔX_{s}, ΔH_{s}, ΔW_{d}, ΔX_{d}, and ΔH_{d} are errors in determining the ground speed, horizontal coordinates, and height, respectively, in the standard and differential (SLS) modes.

For CAS mode 8 (premature descent), the accident rate factor can be estimated as the ratio of height measurement errors using conventional barometric measurements and using SLS exact values. Then, $\mathbf{F}_{\mathbf{Acc}} = \Delta H_{\mathrm{BA}}/\Delta H_{\mathrm{d}} \approx 10\text{M}/1\text{M} = 10.0$, where ΔH_{BA} is the error of the barometric altimeters.

Table 5.5 contains the calculated expert values of the relative flight safety enhancement factors when using SLS data in the CAS and when using CAS data in the SLS.

The analysis of the obtained results shows that the integrated use of these systems provides additional advantages both in case one of the systems fails, and when using data from one system to support the functions of the other one.

The next section will consider recommendations for the practical application of the proposed technical solutions for the implementation of an integrated approach in the joint SLS and CAS use, as well as the results of flight and operational testing of these systems.

Table 5.5 Calculated values of the relative flight safety enhancement factors

	F_A	F_{Acc}	F_{Erg}	F_{Cont}	F_{Int}
CAS mode					
Mode 1			17.7		
Mode 2			10.0		
Mode 5	4.0		4.0		
Mode 7		1.7			
Mode 8		10.0			
SLS mode					
Selection of runway, display of deviations from the glide path and from the intended landing point		6.7	2.4	1.4	2.0
Information received when moving along the runway		5.0	2.4	1.4	3.4

5.5 Conclusions

This chapter presents methods for the integrated use of satellite-based landing systems (SLSs) and collision avoidance systems (CASs) based on the complementary functional elements of these systems. This expands the technical capabilities of both the SLS and the CAS and, ultimately, increases flight safety.

Improvement of flight safety based on the method of integrated use of SLS functional elements to expand the CAS capabilities is due to the use of high-precision and reliable SLS navigation information in the implementation of CAS modes 1, 2, 5, 7, and 8. The application of the developed method is characterized by the factor of increased accuracy of $\mathbf{F_A} = 4.0$ (for mode 5), factor of accident rate reduction: $\mathbf{F_{Acc}} = 1.7$ (for mode 7) and $\mathbf{F_{Acc}} = 10.0$ (for mode 8); factors of increased ergonomics: $\mathbf{F_{Erg}} = 17.7$ (for mode 1), $\mathbf{F_{Erg}} = 10.0$ (for mode 2), $\mathbf{F_{Erg}} = 4.0$ (for mode 5).

Improvement of flight safety during approaches based on the method of integrated use of SLS functional elements to enhance the CAS capabilities includes preventing aircraft landings on an unauthorized runway (RWY) by calculating a virtual glide path, determining, and displaying aircraft deviations from it, and calculating the deviation from the predicted landing point. The application of the developed method is characterized by the reduction in the probability of accidents $\mathbf{F_{Acc}} = 6.7$; ergonomics factor $\mathbf{F_{Erg}} = 2.4$; continuity factor $\mathbf{F_{Cont}} = 1.4$ and integration factor $\mathbf{F_{Int}} = 2.0$.

To improve the safety during the movement after the landing, a complex method has been developed that differs in that after the landing, the aircraft deviations from the runway centerline are calculated and displayed in a graphical form similar to the deviation from the glide path during the approach, and the vectors of the actual and required accelerations are calculated and displayed, as well as the calculated stop point. The application of the developed method is characterized by the accident rate reduction factor $\mathbf{F_{Acc}} = 5.0$; ergonomics factor $\mathbf{F_{Erg}} = 2.4$; continuity factor $\mathbf{F_{Cont}} =$ 1.4 and integration factor $\mathbf{F_{Int}} = 3.4$.

References

1. Baburov VI, Galperin TB, Gerchikov AG, Ivantsevich NV, Sayuta OI, Sokolov AI, Chistyakova SS, Jurchenko YuS (2012) Complex method of aircraft navigation. Application No. 2012136399, priority of 17.08.2012 (in Russian)
2. Baburov VI, Volchok YuG, Galperin TB, Gubkin SV, Maslov AV, Sauta OI, Rogova AA (2011) Method of notification of the aircraft location relative to the runways during the approach and roll-on operation, In: Positive decision on granting a patent following the application for an invention No. 2011113706/11 (020353), priority of 04/04/2011 (in Russian)
3. Gomin SV, Makelnikov AA, Maslov AV, Priyemysheva AA, Sauta OI, Semenov GO, Sobolev SP (2005) Program for warning and emergency alarms generation, Certificate of official registration of computer programs. RF №2005611918. Federal Service for Intellectual Property, Patents and Trademarks (in Russian)
4. AVIATION REGULATIONS (2005) Part 25. Airworthiness requirements for transport aircraft. In: Interstate Aviation Committee. OJSC "AVIAIZDAT", Moscow (in Russian)
5. AVIATION REGULATIONS (1995) Part 29. Airworthiness requirements for transport rotorcraft. In: Interstate Aviation Committee. OJSC "AVIAIZDAT", Moscow (in Russian)
6. Sauta OI, Gubkin SV (1987) Study of the mathematical model of the SRNS onboard receiver input signal in the landing area. In: Radioelectronics issues. Series OVR-1987-Issue 5, pp 39–48 (in Russian)
7. US Patent No. 7 079 951 B2, cl. G01S 13/00, application of 10.12.2004., publ. on 18.07.2006
8. US Patent No. 7 206 698B2, Cl. G06F 17/00, application of 10.12.2004., publ. on 17.04.2007
9. US Patent No. 7 363 145 B2, cl. G08G 5/00, application of 10.12.2004., publ. on 22.04.2008
10. Annual report of the Interstate Aviation Committee (IAC) "Flight Safety Status in 2011" [Electronic resource], www.mak.ru (in Russian)
11. Kotik MG (1984) Aircraft takeoff and landing dynamics. Mashinostroenie, Moscow (in Russian)
12. Qualification requirements "GNSS/SBAS onboard equipment" QE-229. Revision 1. Interstate Aviation Committee-2011 (in Russian)
13. Russian Patent No. 2 312 787, cl. B64D 45/04, application of 25.11.2004, publ. on 20.12.2007 (in Russian)
14. Baburov VI, Ivantsevich NV, Sauta OI (2011) Formalized approach to the selection of a receiver for onboard satellite navigation and landing equipment. In: Radioelectronics issues. Series OT-2011-Issue 4, pp 16–31 (in Russian)
15. Sozinov PA (2011) Applied science and organization of production: [monograph]. In: Rogova AA, Sauta OI (eds) by Doctor of Technical Sciences. Ministry of Education and Science of the Russian Federation, GUAP, St. Petersburg, pp 281–306 (in Russian)
16. Minimum Performance Standards—Airborne Ground Proximity Warning Equipment. RTCA DO-161A, 1976
17. Joint requirements to the airworthiness of civil transport aircraft of the CMEA member countries (ENLG-C). IAC for the airworthiness of civil aircraft and helicopters of the USSR, 1985, 470 p (in Russian)
18. Annex 10 to the Convention on International Civil Aviation. Aeronautical Telecommunications. Radio Navigation Aids, 6th edn, vol 1, July 2006 (in Russian)
19. Scott WB (1996) New technology, training target CFIT losses, aviation week and space technology, 4 Nov 1996, pp 73–77
20. Instruction manual. Enhanced ground proximity warning system. EGPWS. RShPI.461531.001 RE. VNIIRA, 2004 (in Russian)
21. Babich OA (ed) (1981) Aviation instruments and navigation systems. Approved by the commander-in-chief of the Air Force as a textbook for students of Air Force engineering universities: Zhukovsky Air Force Engineering Academy, 348 p (in Russian)

22. Standards for Processing Aeronautical Data. RTCA DO-200A. Radio Technical Commission for Aeronautics, 2005
23. Belyaevsky LS, Novikov VS, Olyanyuk PV (1982) Fundamentals of radio navigation: a textbook for civil aviation universities. Transport, Moscow, 288 p (in Russian)
24. Cherny BF (1972) Propagation of radio waves, 2nd edn. Sov. radio, Moscow, 464 p (in Russian)

Chapter 6
Recommendations for the Application of the Proposed Technical Solutions in the Satellite-Based Landing Systems and Collision Avoidance Systems

This chapter describes the principles of constructing aviation onboard navigation and information systems using GNSS technologies, their functions, and technical characteristics on examples of specific equipment samples.

The formulated recommendations are aimed at improving flight safety and at accelerating the implementation of the basic principles of the aviation equipment development in the regulatory and advisory documents of the International Civil Aviation Organization (ICAO) and other regional and state aviation organizations. These recommendations are described in more detail in [1–7].

As examples of aviation equipment using GNSS technologies and the integrated use of functional modules of complementary systems, three types of products are described that were developed, certified, and manufactured in Russia and perform the functions of satellite-based landing systems (SLSs) and collision avoidance systems (CASs) [8–10].

The onboard navigation and landing complexes (ONLCs) constructed with the use of these products ensure the solution of navigation and landing tasks for flights in accordance with modern principles of area navigation (RNAV) and performance-based navigation (PBN) [11].

During the development and practical implementation of the products described below, the methods, techniques, and devices presented in this paper were used and implemented to improve flight safety [12–22].

Over the past ten years, none of the aircraft equipped with these products implementing the methods, techniques, and devices developed in the present paper have been involved in any accidents.

© Springer Nature Singapore Pte Ltd. 2020

Baburov S.V. et al., *Development of Navigation Technology for Flight Safety*,
Springer Aerospace Technology, https://doi.org/10.1007/978-981-13-8375-5_6

6.1 Principles of Construction and Design Features of Onboard Equipment to Improve Flight Efficiency and Safety

As noted in [23], at present, *three* main directions are singled out that significantly influence the development of new technologies in avionics.

The *first* direction is associated with the introduction of achievements in microelectronics and computer technology. Modern microelectronics makes it possible to deliver practically any circuitry solutions, and the computational resources allow implementing modern methods of optimal signal and data processing, including those based on combining various onboard sensors and systems.

The *second* direction of the development of new technologies is associated with the improvement of conventional technologies and the creation of new systems that complement and expand each other's capabilities. The new direction of avionics development in civil aviation is defined by the CNS/ATM concept (communication, navigation, surveillance/air traffic management) developed and implemented by ICAO. The core of the CNS/ATM concept and systems is the GNSS standardized by ICAO and supported by the military departments of the USA and Russia. For the most effective use of civil aviation, ICAO and ARINC have developed the area navigation concept (RNAV) and the required navigation performance (RNP) concept, which, in recent years, have evolved into the concept of performance-based navigation (PBN) [24]. All these concepts are based on the requirements for accuracy, integrity, continuity of service, and availability which are currently provided through the creation of GNSS augmentation systems discussed earlier in Chap. 2.

Finally, the *third* main direction in the development of aviation technology is the creation of new principles for the organization and interaction of various systems. Here, the use of multi-functional integrated onboard equipment is meant: computer and radioelectronic equipment means for information transmission and display. The integration is one of the fundamental trends in modern avionics. In pursuing the goal of a comprehensive solution to the tasks of navigation, landing, electronic warfare, and weapon control, numerous studies on avionics development programs have been carried out in several leading countries of the world. In evaluating avionics as a unified technical field for providing air navigation, integration of various means and subsystems is always defined as the main direction of its improvement.

Currently, the avionics integration process continues and expands.

As noted in [23], there are virtually no clear ideas about functional tasks, principles of organization and functioning of the integrated onboard radioelectronic equipment (avionics).

We will try to consider the construction of an integrated system (avionics) on the basis on the onboard equipment of the satellite-based landing system (SLS) and the collision avoidance system (CAS) from common methodological positions based on the use of GNSS technologies.

Figure 6.1 summarizes the modern technology of constructing any avionics systems.

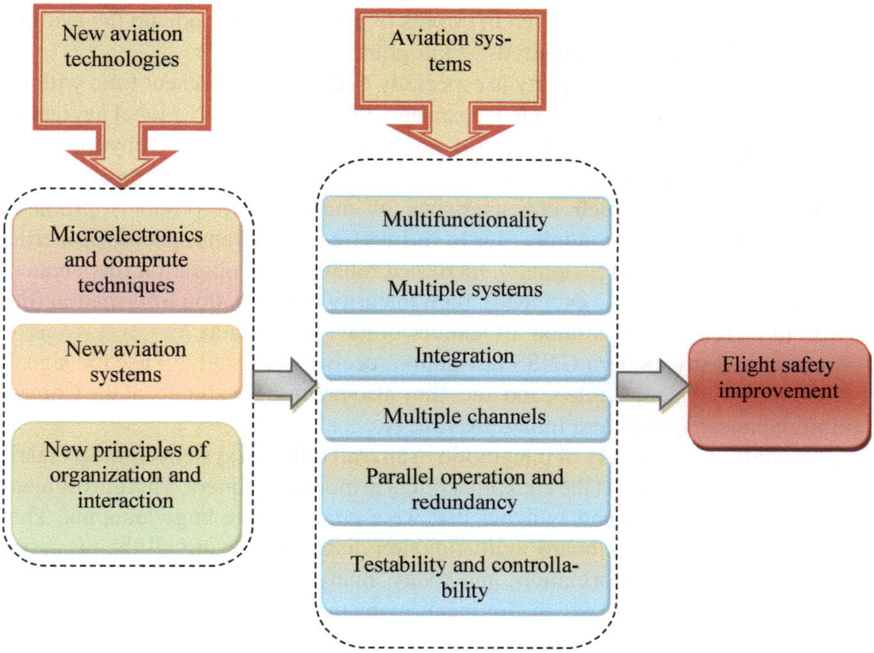

Fig. 6.1 Technology for constructing avionics systems to improve flight safety

At the core of the integration under consideration, there are several prerequisites of a technical nature. Integration of systems essentially different in terms of functional tasks makes it possible to demonstrate its main advantages most clearly.

Multi-functionality. The property of multi-functionality is characteristic for the development of all technical systems for both industrial and domestic uses. This property is now inherent in most technical devices. Even such a utilitarian device as a telephone has turned, in recent years, into a multi-functional complex that performs a wide range of tasks with special applications, from the camera to the navigation system.

The property of multi-functionality at all stages of the development of technical systems made it possible to provide both an increase in the efficiency of solving specific problems for which the device was actually created and increased its competitiveness.

Currently, the avionics structure consists of many multi-functional devices. These include: multi-functional displays and onboard computer systems, navigation and landing complexes and multi-functional communication systems, and systems for preventing aircraft collisions in air and with ground.

The process of further integration of multi-functional systems continues, and there are already onboard avionics models that are capable of solving problems of a whole onboard radioelectronic complex in one relatively small unit of equipment;

and decades ago, these were solved in dozens of products that made up the avionics complex weighing tens and hundreds of kilograms.

Multiple systems. This property is especially evident at the present time with the use of GNSS technologies. Just a few years ago, GPS was the only global navigation system widely known all over the world, and now no one is surprised by the multi-system GLONASS/GPS receivers that have significantly higher user characteristics.

For aviation users, such characteristics of multi-system products primarily include: increased noise immunity (due to various frequencies and spectral properties of the GLONASS and GPS signals), increased reliability of operation (dependence on failures in one system is excluded), high integrity of navigation information (due to twofold increase in the number of sources of navigation data).

Moreover, many modern GNSS navigation receivers are ready to use the signals of the European system Galileo, and later they also process the signals of the Chinese Beidou, Japanese QZSS, and Indian IRNSS, etc.

Integration. The property of integration in aircraft onboard systems is most clearly manifested when looking at the cockpit of pilots in modern airliners. There are already no instruments, buttons, and switches that were previously in huge amounts. They are replaced by several onboard multi-function displays showing all the necessary navigation and flight information, and today, many displays already support the functions of touch control, which eliminates many mechanical buttons and switches.

The same screen displays both flight information, and synthesized aeronautical data, and information from television systems, thermal imagers, radars, etc.

A similar situation is observed in the construction of transceivers. Today, in the same receiver, navigation data can now be extracted from radiotechnical signals by digital processing, separation, and filtration. And from here follows the next logical step, an integrated multi-band antenna that replaces the whole antenna-feed system, usually significantly complicating the design of the aircraft and its operation in general.

Multiple channels. This property is the most inherent to the equipment in which GNSS technologies are used. Indeed, it is hard to imagine now that a few years ago single-channel GNSS receivers were used in which the necessary data from several NSVs were obtained by multiplexing, i.e., alternating reception of signals from several NSVs (usually from four ones, providing the best geometric factor). This, of course, led to additional navigation errors and did not contribute to improving the reliability of aviation equipment, which is extremely important for flight safety.

At present, GNSS receivers have already been manufactured and used in aviation that has several hundred parallel channels for the reception of radio navigation signals. For example, in a TR-G3 [25] module, up to 216 channels are used. This number seems redundant, but in fact, in the near future receivers will appear with even more number of channels! What such a number of channels provide? Firstly, it becomes possible to optimally process signals from all visible NSVs. Currently, their number can reach 30 using the GLONASS and GPS systems, and with the advent of the Galileo system in the near future, this number will come close to 40–50. Using such a high redundancy of information (to determine the position in aviation, at least five NSVs are needed), it is possible to provide the highest requirements to the accuracy of determining the aircraft coordinates and speed. Secondly, high information redundancy makes it possible to meet the highest requirements to the integrity of navigation information, which is one of the determining factors in using such critical applications as, for example, aircraft landing systems with satellite technologies. Thirdly, the use of multiple channels makes it possible to accelerate the search for signals many times and, consequently, to increase the signal continuity and availability of the equipment.

When using multi-frequency GNSS receivers, the number of required channels for signal reception increases several times in comparison with the number of channels mentioned above. In the coming years, new GNSSs will appear and, naturally, new channels will be needed to receive navigation signals from them. Additional channels are also required for receiving information from GNSS augmentation systems, for example, from satellite-based augmentation systems (SBASs), and for access to authorized channels, the use of which is planned in the Galileo system.

Thus, the use of multiple channels is one of the fundamental principles that is used in modern GNSS aviation technologies.

Parallel operation and redundancy. In accordance with aviation regulations [26–28], at least two radio navigation systems must be installed onboard the aircraft. This is due to both safety requirements and requirements for the reliability and integrity of navigation information.

The use of two or more simultaneous sets of onboard equipment ensures the improvement of the continuity and reliability of navigation support at all phases of the flight. For example, in modern aircraft, if one of the navigation equipment sets fails, the system automatically switches to the other set. The availability of navigation parameters formed generated by two or more sets simultaneously makes it possible to solve a number of tasks to ensure the reliability, continuity, and integrity of navigation information.

More challenging is the task of providing parallel operation of several sets of the equipment, information from which is used either by pilots of the aircraft for manual piloting or is sent directly to the aircraft automatic control system. For example, it is unacceptable when different aeronautical databases are used in different sets of computing devices generating a desired flight path, or the pilot manually corrected the flight plan in one of the devices, but this information was not received in the other device. To avoid such a situation, special hardware and software are provided in the onboard equipment, in particular, in the GNSS equipment, which ensure the synchronization of the operation of several parallel sets of the equipment.

Testability and controllability. For aviation applications, at present, the tasks of equipment testability and controllability assume great importance.

The high level of the GNSS technologies forms in pilots a high degree of confidence in navigation information coming from the GNSS. However, in any technically complex system, situations are possible where the information provided is unreliable.

In order to avoid such situations, special methods of data integrity monitoring are used in the onboard equipment that are based on the use of information from other onboard sensors and systems. Such methods have received the general name—aircraft-based augmentation system (ABAS). ABAS complements and/or integrates information received from the GNSS elements with information available onboard the aircraft in order to ensure the operation in accordance with the requirements for GNSS accuracy, availability, and reliability.

First of all, ABAS provides integrity monitoring to solve a navigation task using redundant information (for example, multiple range measurements). The monitoring generally consists of two functions: error detection and error exclusion. The purpose of error detection is to detect false positioning. When it is detected, the source of the error is properly identified and excluded (identification of an individual source creating the problem is not necessary), thereby ensuring continuous GNSS-based navigation.

There are two main classes of integrity monitoring: receiver autonomous integrity monitoring (RAIM) that uses only the GNSS information, and aircraft autonomous integrity monitoring (AAIM) that uses information from additional onboard sensors (e.g., barometric altimeters, clocks and inertial navigation systems);

The ABAS provides continuity for the navigation solution by using information from alternative sources, such as the inertial system, the barometric altimeter, and the external clock. In addition, ABAS maintains availability when solving a navigation task (similar to maintaining continuity) and maintains the accuracy of onboard aids by estimating the residual errors in measured ranges.

Information from other sources can be combined with the GNSS information in two ways:

(a) through an algorithm integrated within the navigation GNSS task (an example is modeling of altimeter data as an additional measurement from a conditional satellite);
(b) using an external (relative to the GNSS) position (an example is the comparison of altimeter data with the navigation solution with the vertical component and generation of alarm signals when this comparison reveals a large discrepancy).

Each monitoring method has its own specific benefits and disadvantages, so it is not possible to provide a description of all potential integration options with the definition of specific numerical values of the provided performance characteristics. This applies to the situation when several GNSS elements are combined (for example, GPS and GLONASS elements). Such methods are called AAIM (autonomous integrity monitoring).

The controllability of the avionics equipment implies the availability in the equipment of functional capabilities for the evaluation of its operability at any time and under various conditions. Many modern aircraft can be stationed in hangars between flights, where there are no electronic (radio technical) signals of the GNSS, and the preflight operations can be performed while the aircraft are in the hangar. In such situations, the built-in GNSS receivers often include signal simulators that significantly increase the controllability monitoring.

The technology of building aviation systems considered above is used in the structure of specific onboard radioelectronic complexes (ORECs).

The next section gives an example of the OREC structure implemented as part of the Yak-42 aircraft upgrade to refine the approach modes using the SLS and integration of its functions with the CAS functions.

6.2 Construction of an Onboard Navigation and Landing Complex on the Basis of the Satellite-Based Landing System and the Collision Avoidance System

Figure 6.2 shows the structure of the onboard navigation and landing complex of the Yak-42 aircraft, which includes the SLS and the CAS.

The SLS composition includes: TAWS unit and OMS-Indicator. The SLS structure includes DDRE units (two sets, left and right), the AFS simultaneously performing the functions of the ILS antenna system with two half-sets of the onboard equipment "Kurs-MP-70" (left and right) and two flight navigation instruments (FNIs) of the left and right pilots, as well as control panels (PU-1P) and switching devices (PK-42, UK NCH). Buttons "SNS Approach" for the right and left pilots provide, at the choice of the operator, the use of the approach data from the ILS or SLS system, which is provided by connecting high-frequency signals from the AFS either to the ILS receiver (Kurs-MP-70) or to the VDB signal receiver (DDRE).

The features of the onboard complex presented in Fig. 6.2 are:

- capability to perform an approach using both an ILS-like landing system and an SLS;
- single AFS to provide instrument approaches and landings using the ILS system or the SLS;
- two sets of the onboard SLS equipment operating in parallel (DDRE, OMS-Indicator);
- use of GNSS data in the CAS from the SLS, the GNSS receiver of which is capable of providing navigation information in both differential and standard modes of operation;
- single display for the navigation and landing information and the underlying surface.

The onboard SLS and CAS equipment in the hardware compartment and in the cockpit of the Yak-42D aircraft is shown in Fig. 6.3.

An example of the ground-based SLS equipment (LAAS-A-2000) at the aerodrome is shown in Fig. 6.4.

A feature of the ground-based SLS subsystem located on the aerodrome is the extended circular pattern of the VDB transmitter antenna, which ensures reception of a signal onboard the aircraft at any point within up to 50 km from the aerodrome, and SLS operation when the aircraft is moving along the runway.

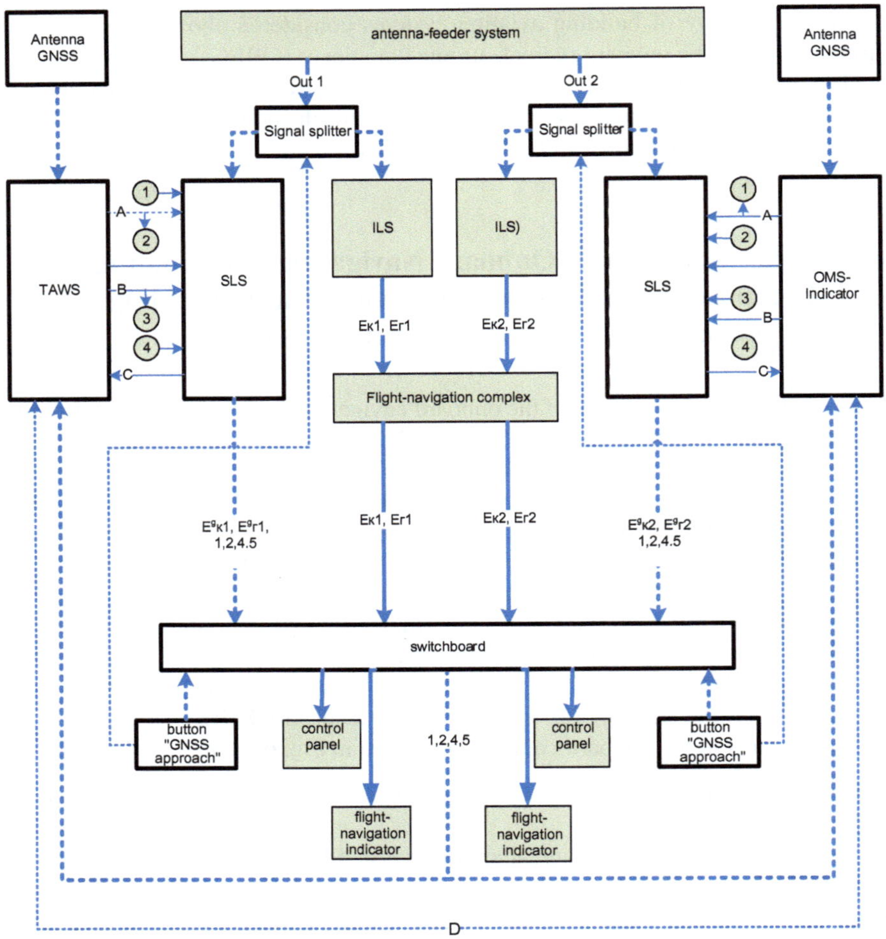

Information in communication channels:

A —number of active channel of GBAS.
B —latitude, longitude, altitude in differential mode.
C —number of active channel of GBAS, angular and linear deflections from glide path.
D—nominal communication channels.

Fig. 6.2 Structure of the onboard navigation and landing complex of the Yak-42 aircraft using the SLS and CAS

a. DDRE

b. CAS equipment (EGPWS unit)

c. CAS display in the cockpit

d. Aircraft Yak-42D on the ground

Fig. 6.3 Onboard SLS and CAS equipment on the Yak-42 aircraft

a. VDB transmitter antenna

b. LAAS-A-2000

Fig. 6.4 LAAS-A-2000 at Ostafyevo airport

The onboard complex designed and built in accordance with Fig. 6.2 is, in fact, an integrated system, which is included in the general structure of the onboard navigation and landing complex of the Yak-42 aircraft.

The ONLC capability to operate simultaneously with ILS and SLS made it possible to perform tests to determine the comparative characteristics of the guidance signals (deviations from the intended glide path) that were generated onboard using the regular airfield ILS and the SLS, the ground subsystem of which (LAAS-A-2000) was also installed for the first time at the airport control tower at a distance of about 1 km from the runway.

During the tests, the guidance signals from the ILS and SLS used for piloting during the approach operations were sent simultaneously to the onboard flight navigation instruments (FNIs) of the Yak-42 aircraft and to the CAS display. At the same time, the SLS display could, at the request of the pilot (or in the automatic mode), show the screen of a virtual FSI or observe the underlying surface above which the flight operations were performed.

Other examples of the construction of onboard navigation and landing complexes (ONLCs) using the principles of SLS and CAS integration are presented in Figs. 6.5 and 6.6.

Figure 6.5 shows the ONLC structure for a Mi-8MTV helicopter. The peculiarity of this ONLC is that its OMS has the function of an integrator of all the navigation and landing information available onboard the helicopter.

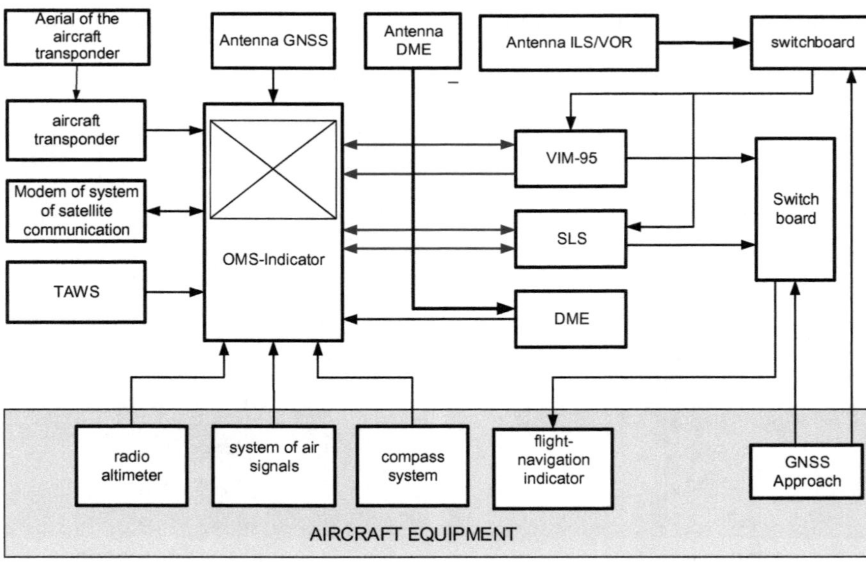

Fig. 6.5 Structure of the onboard navigation and landing complex of the Mi-8MTV helicopter using the SLS and CAS

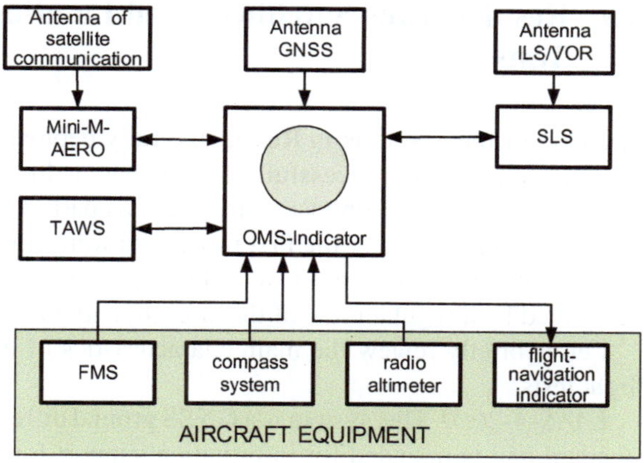

Fig. 6.6 Structure of the onboard navigation and landing complex of the Il-76TD aircraft with integrated SLS and CAS use

Simultaneously, this OMS performs the SLS functions, is a control panel for all systems included in the complex, and also displays navigation information and CAS information (EGPWS).

To ensure the requirements of area navigation, GNSS-based non-precision approaches, GLS-based precision approaches and instrumental monitoring of the aircraft condition in real time, the ONLC of the Il-76TD aircraft was modified.

In November 2007, for the first time in the history of world aviation, an aircraft IL-76TD was landed in Antarctica at the icefield of Novolazarevskaya station in the GLONASS and GPS-based non-precision approach mode using the OMS-Indicator.

Next, we will consider recommendations for the construction of integrated complexes of onboard and radio technical equipment for advanced aircraft, taking into account the methods and approaches developed above (Fig. 6.7).

Fig. 6.7 SLS- and CAS-equipped Il-76TD aircraft in Antarctica at the icefield of Novolazarevskaya station

6.3 Results of the Satellite-Based Landing System Flight Tests

In 2006, for the first time in Russia, the fully deployed and certified satellite-based landing system was successfully tested that included a ground-based subsystem LAAS-A-2000 and an onboard subsystem as part of the DDRE, TAWS, and OMS-Indicator equipment. For the first time, tests were conducted not on a flying laboratory (flight simulator), but on an ordinary passenger aircraft Yak-42D. Flights were performed by a regular crew of the aircraft, and not by test pilots.

Let us briefly review the main characteristics of the equipment used during the flight tests.

LAAS-A-2000. This system is a GNSS ground differential subsystem of the GBAS standard and is intended for information support for ICAO Category I (Categories II and III over time) landings, missed approaches, roll-on operations, run operations and takeoffs, as well as standard procedures of the area (including RNAV, P-RNAV, SID, STAR routes) and en route navigation of the aircraft with the required navigation performance established by ICAO for standard landing and navigation procedures. To ensure its intended use, LAAS-A-2000 (see Table 5.1) generates and transmits messages 1, 2, 4 and 5 over the VDB type data link (DL) in real time in the frequency range of 108.0–136.975 MHz in the ICAO SARPs format to the aircraft and ground users, as well as to ground air traffic control (ATC) services and services for monitoring the GLONASS and GPS satellite constellations via wire, fiber-optic, and satellite data links: differential corrections and their rates of change; data on the final approach segment (FAS) for all ends of all runways within the coverage area of the station; data on the condition and performance of the LAAS-A-2000 equipment and operation modes; real-time information and predicted information on the status of the GLONASS or GPS orbital constellations within the coverage area of the station and other information in accordance with [29], as well as information on integrity violation separately by the GLONASS and GPS constellations and on the LAAS-A-2000 operational status to the ATM system.

LAAS-A-2000 main technical characteristics are given in Table 6.1.

DDRE. The equipment for differential data reception and conversion equipment (DDRE) is the receiving side of the radio technical complex of the ground GNSS augmentation system (GBAS). The ground part of this complex includes a local monitoring and correcting station (LAAS) with a signal transmitter. In fact, DDRE is an onboard receiver of differential data.

The differential corrections, FAS block, integrity parameters, and other data generated and transmitted by the LAAS-A-2000 equipment are received by the onboard differential data receiver (DDRE) and transmitted to the onboard GNSS equipment, which processes GBAS signals and eliminates pseudorange measurement errors in the GNSS receiver for each from the satellites in view.

The DDRE is designed to support precise approaches and categorized landings on runways and sites that do not have any dedicated radiotechnical landing equipment, as well as to support any other applications requiring high-precision positioning.

Table 6.1 LAAS-A-2000 technical characteristics

No.	Name of the characteristic	Units	Value (meaning)
1	Accuracy class for the generation of differential corrections (GAD)		A, B, C, D
2	GNSS systems used		GLONASS, GPS
3	Output data format: ICAO SARPS Annex 10 Volume 1		Message types 1, 2, 4, 5 When using the MLCS for servicing cat. III precise approaches, a message type 2 shall be transmitted
4	*Data update and output interval*		
	1. Differential data 2. Reference station data 3. LAAS ID 4. FAS 5. Predicted availability of the satellites	s s s s s	½ 1 15 15 15
5	Operating frequency of data transmission over a radio channel	MHz	108.00–117.995
6	Center-frequency stability	%	+0.0002
7	VDB transmitter power	W	Up to 150
8	Warm-up time	c	<160
9	*Coverage area for landings*		
	– In the horizontal plane, at least – In the vertical plane, at least Coverage area for RNAV and ADS:	km °	37 7 Line-of-sight, VHF
10	Field strength within the coverage area	dBμV/m	215 min
11	Automatic control system activation when the radiation power is reduced	%	80
12	Time to Alert for the integrity violation	s	<6 (depending on the standard procedure to be serviced)
13	FAS parameters		SARPS ICAO Annex 10, volume 1. message type 4
14	Power supply	V, Hz	380/220 ± 10%, 50 ± 0.1
15	Power consumption	W	900
16	Operating temperature	°C	LAAS +5 to +50
			AFS -40 to +50
18	Time of storage of the registered data	days	14 min

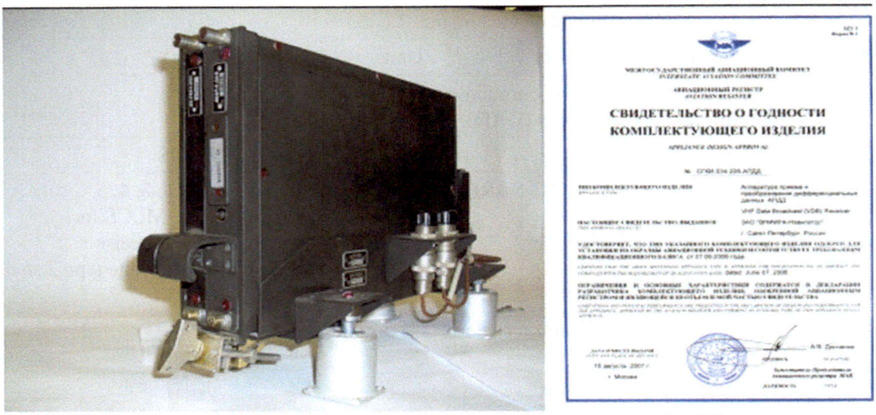

Appearance Certificate

Fig. 6.8 DDRE

The DDRE receives differential data transmitted by the LAAS-A-2000 at an assigned carrier frequency within the frequency band of 108,000–117,975 MHz. The separation between the assigned frequencies is 25 kHz. The dynamic range of the product receiver is at least minus 86 dB. The probability of incorrect reception of the message is such that it is possible to lose no more than one message out of a thousand.

The coding scheme ensures unambiguous assignment of the channel number to each approach with the use of the LAAS-A-2000. The channel number consists of five numeric characters. The channel number allows the receiver DDRE tuning to the correct frequency and selects the FAS data block, which determines the desired approach. The desired FAS data block is selected by the reference path data selector (RPDS), which is included in the message type 4 as part of the data to define the FAS.

The DDRE is operated at ambient temperature from minus 45 to +55 °C and vibrational loads with acceleration up to 20 g. Power consumption over a 27 W network is not more than 30 W; weight is 3.0 kg max, the overall dimensions are as follows: 360 × 200 × 61 mm.

The DDRE appearance and the IAC AR certificate are shown in Fig. 6.8.

The DDRE is designed to be installed on aircraft in service as equipment that provides the flight navigation instruments with differential data obtained from ground-based LAASs and data on deviation from the calculated landing path.

The DDRE receiver is a radio technical facility consisting of devices for receiving, processing, and converting information received from the LAAS over the VHF DL. The receiver control (tuning to the required channel, activation of the built-in monitoring) is provided in accordance with the specifications ARINC429, ARINC755-2. The onboard receiver has receiving channels in the range of 20,001–39,999, where each channel number corresponds to the one-and-only precision approach.

The DDRE provides data on lateral and vertical deviations for the onboard flight navigation complex of the aircraft regarding the selected FAS path over digital and analog output channels.

OMS-Indicator. The product is an onboard multi-functional system with an integrated indicator that provides the function of assessing the terrain in the direction of flight when operating with the EGPWS. The OMS-Indicator is used for flights in accordance with the instrument flight rules (IFR) along the intended route, in the terminal area, during "non-precision" approach (NPA) and, when used together with the DDRE, to ensure ICAO Category I precise landings. The OMS-Indicator is intended for equipping all types of aircraft and solves the following tasks:

- flight planning and execution of en route functions at all flight phases when used as the onboard satellite navigation equipment (OSNE) of class A1;
- receiving and processing navigation signals of satellite navigation systems GLONASS and GPS, providing the autonomous integrity monitoring (RAIM) function, generation and delivery of the corresponding alert/warning and signal information;
- ensuring the input, storage, and use of the aeronautical and operational databases;
- viewing navigation data and indicating the validity of the aeronautical database, and also indicating the validity of the onboard database (ODB) of the Enhanced Ground Proximity Warning System (EGPWS), if the OMS-Indicator interacts with the EGPWS;
- updating the aeronautical database in accordance with aeronautical information regulation and control cycles (AIRACs);
- providing the crew with information on the terrain in the direction of flight, taking into account ground obstacles, textual information on generated signals, the EGPWS status and the associated equipment, if the OMS-Indicator interacts with the EGPWS.
- the OMS-Indicator provides operational control of the type of indication and execution of the aircraft (AC) flight plan with the help of its own controls.

The OMS-Indicator contains the world aeronautical database with standard departure (SID) and arrival (STAR) procedures, provides the generation of navigation parameters necessary for flight along the route and in the terminal area, supports "straight to" flights, flights along parallel routes, flights with a required heading, and flight procedures in the holding area as well.

To determine the current position of the aircraft, the OMS-Indicator provides reception and processing of navigation signals of the GLONASS/GPS satellite navigation systems and performs autonomous integrity monitoring (RAIM function) for navigation data.

The aeronautical database of the OMS-Indicator contains data on aerodromes, runway ends, SIDs, STARs, and non-precision approach procedures; data on navigation points (NPs), including VOR, DME, SRNS and location beacons, data on airways, holding areas, and communication frequencies at aerodromes.

The OMS-Indicator receives input signals from aircraft gear compression sensors, barometric altitude sensors or the air data system, the heading system, the EGPWS, other onboard satellite navigation equipment (including EGPWSs), and other products.

When using a dual set of the OMS-Indicator, the interaction of sets is provided via the ARINC 646 (Ethernet) communication line. This ensures synchronization of the following data:

– flight plan, including inbound, outbound, and approach procedures;
– "straight to" mode;
– flight along a parallel route;
– tuning data ("direct to" flying points, speed, roll, GAMA parameter);
– aircraft position determination mode (constellation of satellites used).

When using a dual set of the OMS-Indicator, it is possible to maintain routes, navigation points, aerodromes, and runways simultaneously in the user database of each of the half-sets.

The main technical characteristics of the OMS-Indicator are given in Table 6.2.

The OMS-Indicator appearance and the IAC AR certificate are shown in Fig. 6.9.

Table 6.2 Main technical characteristics of the OMS-indicator

Name	Characteristics
GNSS operating constellation (band)	GLONASS/GPS (L1)
Coordinates (speed) errors, max	6 m (5 cm/s)
Input/output/two-way channels	8/4/2
Operating temperature	−40 to +55 °C
Warm-up time	3 min max
Power supply	27 V, 35 W
Dimensions	146 × 125 × 225 mm
Weight	3.0 kg

Appearance

СВИДЕТЕЛЬСТВО О ГОДНОСТИ
КОМПЛЕКТУЮЩЕГО ИЗДЕЛИЯ

Multi Function Indication System (MFIS)

dated July 13, 2007

Certificate

Fig. 6.9 OMS-Indicator

In addition to the controls on the front panel, there is a connector over the screen for the database loader connection and a photosensor for automatic adjustment of the brightness.

Multi-functional buttons (MFBs) are located to the left and to the right of the OMS-Indicator screen.

When a message is generated by the Product, the letters "MSG" and the message text are displayed on the bottom line of the screen. If there are several messages, they are displayed alternately, for 1 s each. If the integrity of the GNSS information is violated, the RAIM function is activated and the RAIM indication is displayed.

Let us briefly review the results of the SLS tests with ground and onboard equipment described above that implements part of the methods for improving flight safety discussed in Chap. 3. In the process of testing at the aerodrome, two independent landing systems operated simultaneously: the meter-wave instrument system of type ILS and SLS, with a ground GBAS subsystem based on the LAAS-A-2000 [30].

Ground tests were performed by towing a Yak-42D aircraft along the runway of the aerodrome using an aviation tow tractor. During the towing process, the displacement of the front wheel of the aircraft from the marking of the runway centerline was monitored. In these ground tests, this displacement did not exceed 0.5 m. During the movement, the bars of the heading plane of the standard FNIs were also monitored for the position displacement on the Yak 42D (see Fig. 6.10) and on the OMS-Indicator as well (see Fig. 6.11). The test procedure consisted in that, when moving along the runway, the pilot alternately displayed information from the ILS and SLS using the "SNS approach" button (see Fig. 6.2).

As a result of the tests, it was found that the bars of the heading plane position on the standard FNIs and on the virtual FMIs of the OMS-Indicator product behave identically and show a zero value of the deviation signal ε_h when the aircraft is on the

Fig. 6.10 Appearance of standard FNIs on a Yak-42D aircraft

Fig. 6.11 Appearance of virtual FNIs on the Indicator

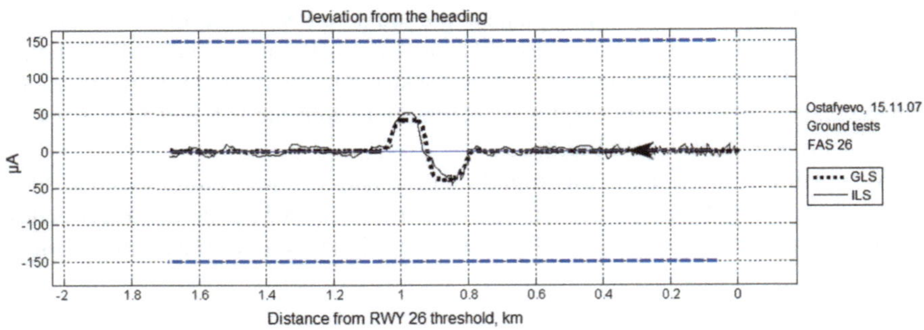

Fig. 6.12 Ground tests with the aircraft moving along the runway

runway centerline and deviate to the same side with the displacement of the aircraft from the centerline.

To determine the SLS sensitivity to the deviation of the aircraft from the runway centerline, an S-shaped maneuver was performed. During the test, SLS and ILS data were simultaneously registered.

Figure 6.12 shows the results of the tests when the aircraft moved along the runway.

The analysis of the results of ground tests presented in Fig. 6.12 shows that the SLS-generated ε_h deviation values have a smaller noise component than the matching values generated by the ILS.

At the second stage, the SLS was tested during approaches in the manual piloting mode. At the same time, a standard Category I ILS landing system was used as reference equipment. A few dozen approaches were performed. Examples of processing the results obtained during these approaches are shown in Fig. 6.13.

Figure 6.13 shows the projections of the aircraft movement trajectory onto the horizontal and vertical planes (see Fig. 6.13A1–F1) (along the horizontal axis the distance to the runway threshold is shown), as well as the corresponding values of deviations ε_h from the heading plane and ε_g from the glide path plane—dashed line by the SLS data, and solid line by the ILS data (see Fig. 6.13A2–F2).

To determine the SLS sensitivity to the aircraft deviation from the heading plane, the approach shown in Fig. 6.13D1, D2 was performed in the "S on the heading" mode, and to determine the SLS sensitivity to the aircraft deviation from the glide path plane, the approach presented in Fig. 6.13E1, E2 was performed in the "S on the glide path" mode.

The analysis of the test results presented in Fig. 6.13 showed that the generated SLS and ILS guidance signals are practically identical.

It was noted in [30] that the error in determining the coordinates in the SLS differential mode was 0.5 m (2σ) in the horizontal plane and 0.9 m (2σ) in height.

When performing these approaches, the deviation of the heading position and glide path bars was monitoring continuously, without fluctuations and ensured satisfactory piloting in the hand (wheel) control mode.

SLS-based approaches using the onboard DDRE and OMS-Indicator equipment in the manual (wheel) piloting mode are ensured up to a height of 50–60 m.

The piloting of the aircraft following the FNI position bars during the SLS-based approach is practically the same as for ILS-based approach and does not cause any difficulties.

Next, we consider the results of the CAS tests and operation using the example of an Enhanced Ground Proximity Warning System (EGPWS).

6.4 Results of the Enhanced Ground Proximity Warning System Flight Tests and Operation

CASs with the enhanced ground proximity warning mode were developed and first introduced into operation onboard the Russian-made aircraft in 2002. The first Russian CASs were developed in St. Petersburg and upon receipt of the Certificate [31] their commission began—first for civil aircraft, and then for state aviation aircraft.

To date, the EGPWS has been installed and successfully operated by more than 500 aircraft of the following types: An-26, An-70, An-72, An-74, An-140, Be-200, Il-62, Il- 76, Tu 134, Tu-154, Tu-204, Tu-214, Tu-334, Yak-40, Yak-42 and helicopters: Ka-226, Ka-32, Mi-8, Mi-17, Mi-26. The ONLC of many of these aircraft also includes an SLS on the basis of the DDRE and OMS-Indicator, which together with the EGPWS form an onboard navigation and landing integrated complex.

Availability of the TAWS onboard the aircraft is currently a mandatory condition for aircraft approval for any types of flight. At the same time, the EGPWS can

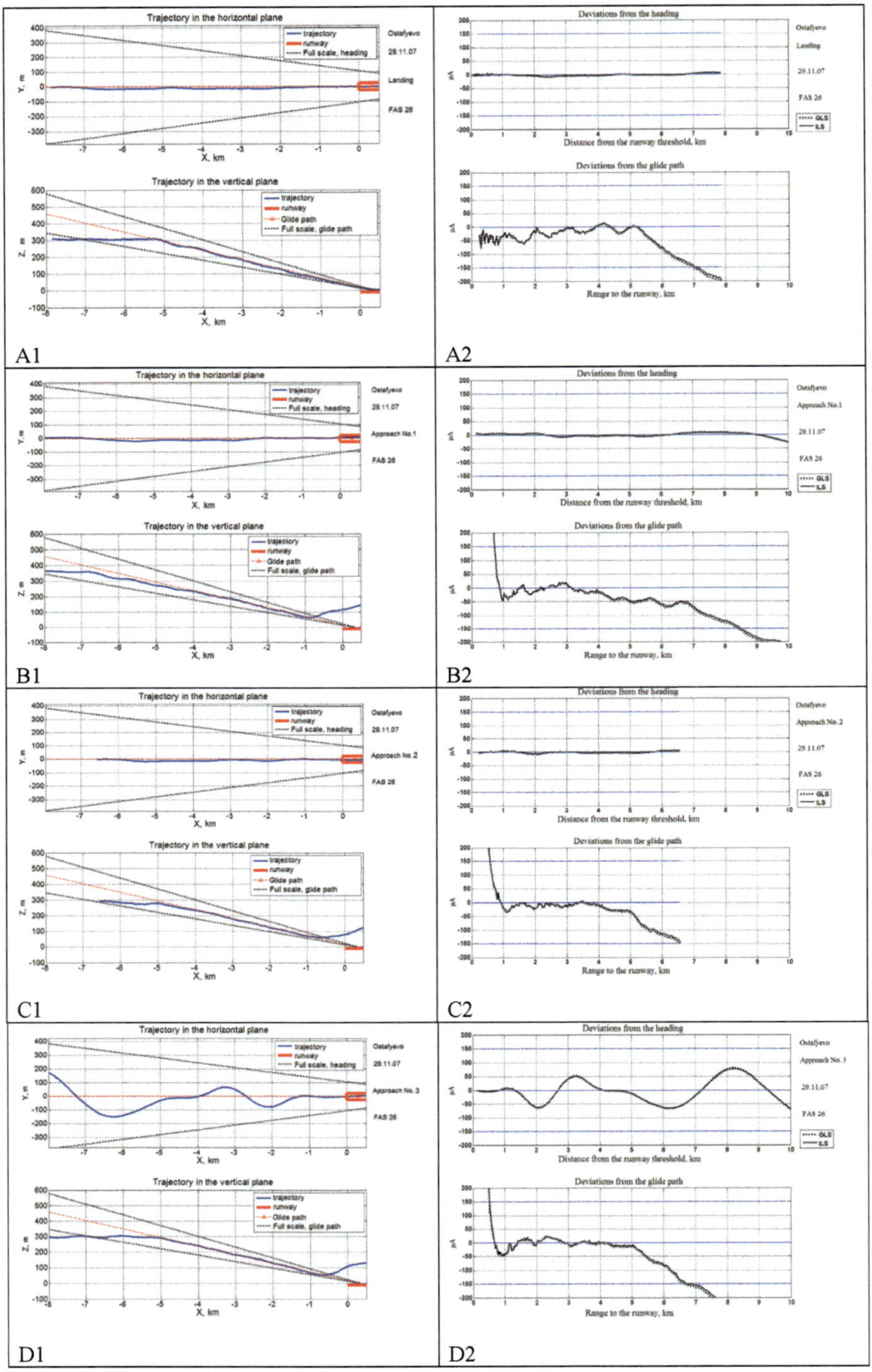

Fig. 6.13 Approaches at the Ostafyevo airport using the SLS

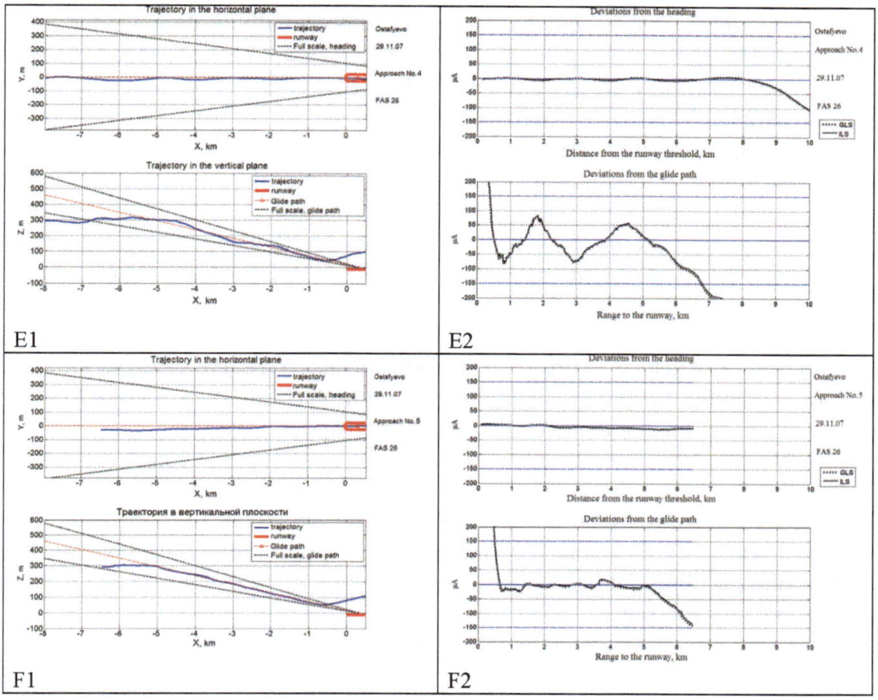

Fig. 6.13 (continued)

also function as a backup (redundant) GNSS information sensor, which increases the integrity and continuity of the onboard complex including the OMS-Indicator, DDRE, and EGPWS.

The EGPWS is a means of improving flight safety by providing, with the help of audio and visual signals, alerts/warnings for the crew of the occurrence of such flight conditions, the development of which can lead to an aircraft collision with the ground or water, as well as with artificial obstacles.

The EGPWS is designed to generate and provide the crew with signals about hazardous proximity of the aircraft and the earth's surface, to transmit the information for displaying the terrain in the direction of flight to the onboard information indicator, taking into account ground obstacles, textual information about signals generated, the EGPWS status and the interacting equipment.

Alerts/warnings are generated by the TAWS at values of the current flight height below the established minimum allowable values or for deviations downward from the equisignal zone of the radio technical glide path (RGP) exceeding the maximum allowable values. Early warning signals are generated when the aircraft approaches the terrain or artificial obstacles, which present a potential danger of collision. Alerts/warnings by the EGPWS are output continuously until the aircraft crew eliminates the cause of a respective alert/warning.

Minimum allowable heights and distances to potentially hazardous sections of the terrain for various flight phases and conditions are established proceeding from the provision of timely alerts/warnings for the aircraft crew of the potential hazard of collision in most possible situations.

Alarm conditions are automatically adjusted in the EGPWS taking into account the current values of the onboard sensor signals and also depending on the landing gear position.

In order to perform the required functions, the TAWS receives signals from the onboard GNSS receiver (external or integrated/built-in), from the radio altimeter, from the onboard equipment of the instrument landing system (ILS), the barometric altitude sensor or the air data system (ADS), the airspeed sensor, the TAWS indicator, the multi-purpose display (MPD) compatible with the TAWS, the "stall" mode alarm system or another system (systems) having higher priority of announced (voice) messages, the sensors of the landing gear position and the controls.

The TAWS controls are usually included in the aircraft or helicopter equipment.

The TAWS sends signals to the flight intercom (FI) system or directly to the crew headsets, to the cockpit loudspeaker, to the alarm light panels or to the emergency warning system (EWS), to the TAWS indicators or multi-purpose displays (MPDs), to the onboard flight data recorder, to other flight safety systems having a lower priority of voice messages.

The TAWS provides alerts/warnings for the crew in the following situations, which can lead to a collision of the aircraft with ground or an artificial obstacle:

- excessive rate of descent,
- excessive terrain-closure rate,
- insufficient clearance margin over the underlying surface during a flight with a large instrument speed,
- exceeding the permissible deviation below the glide path during the instrument approach,
- exceeding the threshold value of the difference between the geometric and relative barometric heights,
- assessment of the terrain in the direction of flight,
- exceeding the permissible bank angle near the earth's surface,
- exceeding the permissible pitch angle near the earth's surface,
- risk of encountering a ring vortex.

The aircraft position is predicted and the safety of this position is assessed with respect to the terrain and artificial obstacles on the basis of information from the GNSS and barometric altimeter data.

The onboard database (ODB) of the EGPWS consists of a periodically updated terrain relief database (TRD), a database of artificial obstacles and a database of runways at aerodromes.

When the TAWS operates in the mode of "Forward terrain assessment," signals are generated based on the results of the analysis of distances to the earth's surface implementing two functions: the function of ensuring the minimum allowable height

Fig. 6.14 TAWS

Appearance Certificate

(MAH) implemented within 0–45 s of the flight and the function of early ground proximity warning (TAWS) implemented within 30–60 s of the flight.

At all flight phases, the TAWS operates automatically and does not require control actions from the crew. When receiving TAWS alerts/warnings, the crew performs actions in accordance with the requirements of the Aircraft Flight Manual, which ensure the elimination of the cause that triggers these alerts/warnings.

To establish the required TAWS operation mode and to adapt it to the onboard equipment installed on the aircraft, contact pins for programmable one-time commands (event-commands) are used. These pins are programmed when installing the TAWS on the aircraft by installing (or not installing) jumpers between the corresponding contact pins.

The EGPWS appearance and the IAC AR certificate are shown in Fig. 6.14.

EGPWS main technical characteristics are given in Table 6.3 and in [8].

Table 6.3 EGPWS main technical characteristics

Name	Characteristics
GNSS operating constellation (band)	GLONASS/GPS (L1)
Coordinates (speed) errors, max	6 m (5 cm/s)
Analog input channels	8
Digital input/output/two-way channels	11/6/2
Input/output channels for one-time commands	25/17
Output audio channels	2
Mean time between failures	10 000 h min
Power supply	~115 V, 400 Hz, 25 VA and/or 27 V, 25 W
Dimensions	57 × 190 × 220 mm
Weight	2.8 kg

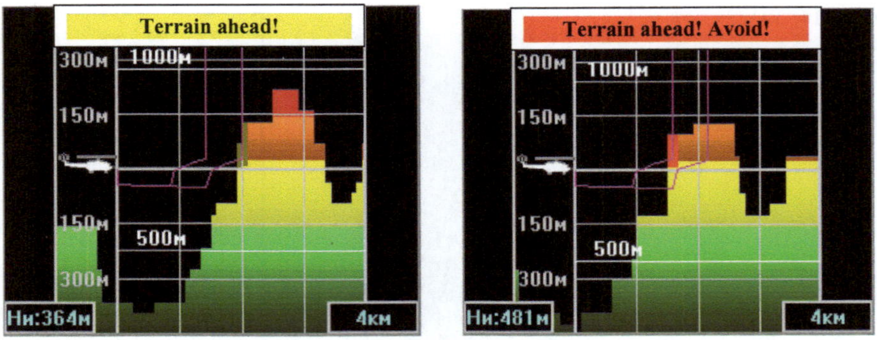

Fig. 6.15 Level flight toward the mountain

To implement its functions, the TAWS provides complex processing of input signals, calculation of the trigger zone boundaries, comparison of the current values of the signal parameters with the calculated values of the trigger zone boundaries, logic processing of signals, generation of output signals, prioritization of voice messages, and built-in testing of operability (health).

The generated voice alerts/warnings are provided for the aircraft pilots through loudspeakers and headsets.

Below are the results of the testing and operation of the TAWS that implements some of the methods for improving flight safety discussed in Chap. 4 of this paper. TAWS tests and operations were performed with the OMS-Indicator or standard onboard multi-purpose displays (MPDs) installed on various types of aircraft. In these tests, certified terrain relief databases included in the EGPWS were used.

Figures 6.15, 6.16 and 6.17 show the results of the TAWS flight tests with the OMS-Indicator installed on a Mi-8T helicopter, under the conditions of flat and mountainous terrain with a relative elevation difference of up to 3,800 m and in flights toward artificial obstacles of up to 250 m over the mountainous terrain on the Kamchatka peninsula.

Figure 6.15 shows the screen of the OMS-Indicator at the moments the EGPWS alarms appeared with the issuance of caution signals (CSs) "Terrain ahead!" (the duration is 11 s) and alarm signals (ASs) "Terrain ahead! Pull up!" (the duration is 30 s). At the time of the CS appearance, the aircraft was 127 m below the TRD element that caused the signal and at a distance of 2.0 km. The flight time to the mountain at a current speed of 143 km/h is 50 s. At the time of the AS appearance, the aircraft was 100 m below the TRD element that caused the signal and at a distance of 1.5 km. The flight time to the mountain at a current speed of 154 km/h is 35 s.

Figure 6.16 shows the screen of the OMS-Indicator at the moments the EGPWS alarms appeared with the issuance of the caution signal "Terrain ahead!" (the duration is 14 s) and the alarm signal "Terrain ahead! Pull up!" (the duration is 31 s) while performing a flight with suspended load. At the time of the CS appearance, the suspended load was 68 m below the TRD element that caused the signal and at a

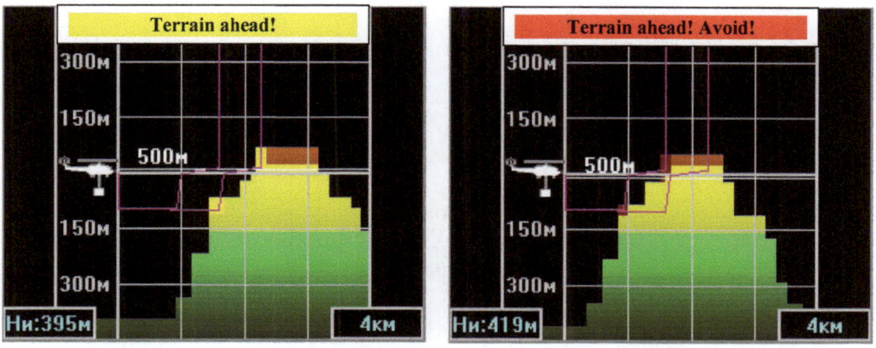

Fig. 6.16 Level flight toward the mountain with the "Suspended load" mode on (50 m long)

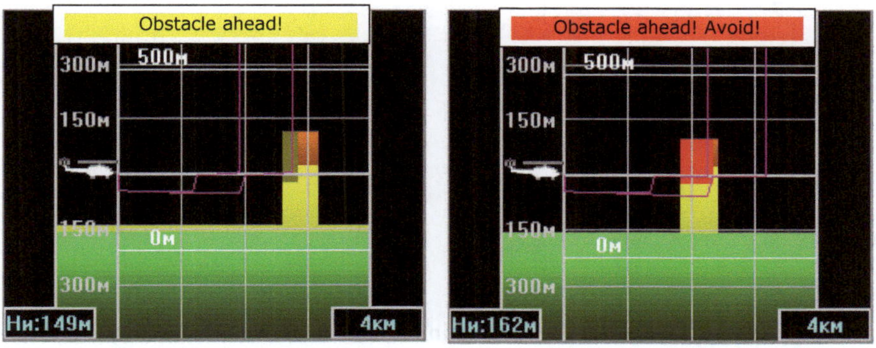

Fig. 6.17 Level flight toward the artificial obstacle

distance of 2.17 km. The flight time to the mountain with a current speed of 159 km/h is 49 s. At the time of the AS appearance, the suspended load was 50 m below the TRD element that caused the signal and at a distance of 1.5 km. The flight time to the mountain with a current speed of 161 km/h is 34 s.

Figure 6.17 shows the screen of the OMS-Indicator at the moments the TAWS alarms appeared with the issuance of the caution signal "Obstacle ahead!" (the duration is 10 s) and the alarm signal "Obstacle ahead! Pull up!" (the duration is 19 s). At the time of the CS appearance, the aircraft was 123 m below the obstacle that caused the signal and at a distance of 2.6 km. The flight time to the obstacle with a current speed of 193 km/h is 48 s. At the time of the AS appearance, the aircraft was 95 m below the TRD element that caused the signal and at a distance of 1.8 km. The flight time to the mountain with a current speed of 216 km/h is 30 s.

Based on the results of these tests, as well as other numerous tests and EGPWS operation on various types of aircraft, it was concluded that it is reasonable to use the TAWS as an "information system for improving the navigation safety."

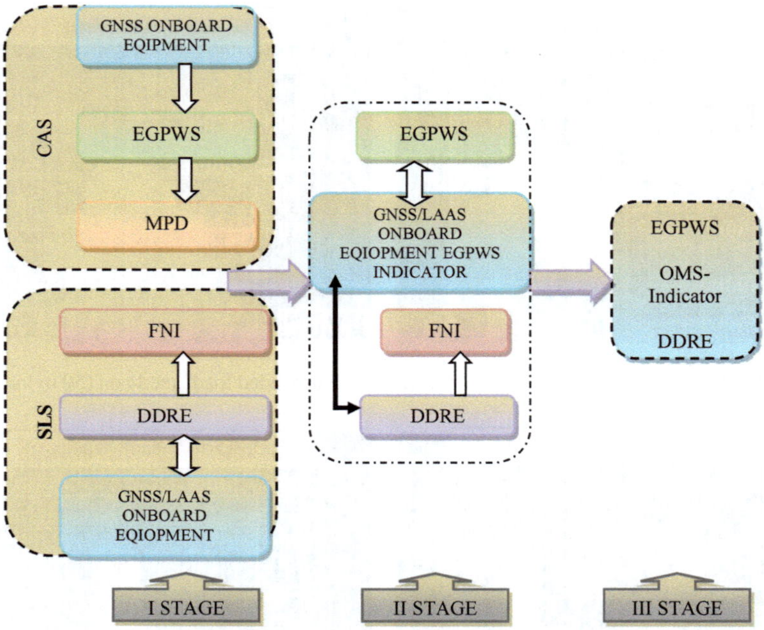

Fig. 6.18 SLS and CAS integration stages

A distinctive TAWS feature is the introduction of the methods and devices described in Chap. 4 of this paper into its software and algorithmic support.

Thus, for example, the types of screens used by pilots presented in Figs. 6.16, 6.17 and 6.18 represent the cross projection of the underlying surface and the corresponding projections of the protection space that the majority of foreign systems of this class do not contain.

For clarity, the presented test results are summarized in Table 6.4.

The analysis of Table 6.4 shows that the presented screen for the OMS-Indicator provides additional time for decision-making by the pilot within 15–18 s after the CS appearance.

The display of the protective space makes it possible to avoid CS appearance, in general, and this time margin can reach several tens of seconds. The results of the flight tests, in general, confirm the expert estimates of the increase in flight safety

Table 6.4 EGPWS signal generation

	Time (s)	Range (km)	Time (s)	Range (km)	Time (s)	Range (km)
CS	50	2.0	49	2.17	48	2.6
AS	35	1.5	34	1.5	30	1.8
Figs.	6.14		6.15		6.16	

obtained in Sect. 6.4 of this paper, according to which the value of the ergonomics factor F_{Erg} lies in the range from 2 to 5 using the proposed methods for increasing flight safety.

In conclusion, it can be said that if we assume an average annual flight time per aircraft of 3500 h and take into account that the TAWS has been operated for more than ten years, the total TAWS flight time on all types of aircraft will be: $(700/2) \cdot 10 \cdot 3500 = 2{,}250{,}000$ h. During this time, there was not a single catastrophe caused by a collision with the ground or an obstacle among EGPWS-equipped aircraft. If we take an average flight time of 2.5 h, this means that 4.9 million accident-free flights have been performed, which corresponds to a safety factor of $\approx 2 \cdot 10^{-7}$.

Let us now turn to an integral evaluation of improving flight safety with the joint SLS and CAS use.

Integral evaluation of the flight safety enhancement and recommendations on the practical application of the proposed technical solutions.

Figure 6.18 shows three stages of integration (performed in the last ten years) of the systems initially independent from the perspective of hardware and functionality, SLS and CAS, but performing one of the most important functions—improving flight safety. At the first stage, in the early 2000s, the two independent systems developed in parallel (see Fig. 6.18). At the second stage, since 2005, obvious interrelations appeared between the systems that, by the end of 2008, led to the third stage of integration, when the systems became so close in respect of hardware that the next step would inevitably be their complete integration.

Figure 6.18 reflects the fact that the process of integrating the systems that are initially not connected but performing one common function of improving flight safety, inevitably leads to their integration into a single system.

At present, the degree of equipment integration is constantly growing due to the rapid development of the element base of analog–digital converters, digital computing devices and other electronic components (ECs). This allows the implementation of functions of many previously independent systems in one relatively small unit of equipment based on the principles described in Sect. 6.1 of this chapter.

In practice, onboard equipment such as "OMS-2010" already exists, in which the above-mentioned OMS-Indicator integrates SLS, ILS/VOR, and VHF radio-station functions.

Thus, a large part of the traditional onboard navigation and landing complex is implemented in a product with a volume of less than 3 dm^3.

Let us perform an integral evaluation of the increase in flight safety on the basis of the indicators defined in Sect. 1.6 using the estimates of the relative flight safety enhancement factors obtained earlier in Chaps. 3, 4, and 5.

The analysis of the results obtained allows drawing the following conclusions:

The use of these methods ensures the improvement of flight safety.

All the methods presented, which are characterized by increased accuracy, integrity, and continuity of the SLS navigation information increase in the $\mathbf{F}_{\mathbf{Acc}}$ factor (that is, reduce the probability of accidents), since they provide a more adequate representation of the actual parameters of the aircraft condition.

The integrated SLS and CAS use provides an additional contribution to improving flight safety, regardless of whether any other methods are applied. That is, in itself, the integrated (simultaneous) SLS and CAS use is an independent method for improving flight safety.

The analysis of the obtained expert estimates of flight efficiency and safety factors makes it possible to select those methods, the implementation of which can have the greatest positive effect when the SLS and CAS are put into operation on particular aircraft and at specific airfields.

Returning to the formalized statement of the problem defined in Sect. 1.6, it can be noted that it is possible to ensure the maximum flight safety ($P_{FS}(t) \rightarrow 1$) with simultaneous use of all the methods described above. At the same time, since each of the methods is independent of the others, it can be expected that the total relative increase in flight safety will be proportional to the sum of the partial increments in accordance with expression (1.11).

It should also be noted that the considered methods and devices for improving flight safety do not exhaust all possible technical approaches to solving the problem of flight safety improvement, but make it possible to significantly improve flight safety during flights under the conditions of hazardous terrain, during approaches and landings, that is, in the most accident-prone situations.

6.5 Conclusions

The main principles of constructing modern systems for improving flight safety are: multi-functionality, multiple systems, integration, multiple channels, parallel operation and redundancy, testability and controllability.

The design features of constructing SLS- and CAS-based systems for improving flight safety are the use of the following common devices: an antenna-feed system, multi-purpose displays, and onboard GNSS receivers that shall be selected using the methodical approaches described in Chap. 2.

The developed structures for the construction of onboard navigation and landing complexes, on account of the methodology of their construction described in Chap. 2, proved effective when implemented on various types of aircraft for technical, temporal, and economic characteristics. The products built on the basis of the use of these structures demonstrated high performance.

The TAWS, OMS-Indicator, and DDRE developed and put into operation currently reflect the highest point of development of the domestic onboard avionics using GLONASS/GNSS technologies. More than 2000 domestic aircraft of state and civil aviation are currently equipped with these products.

To date, aircraft equipped with the EGPWS have performed about 5 million flights, while the flight safety evaluation is five times higher than ICAO requirements and today it is $\approx 2 \times 10^{-7}$.

The performed full-scale tests confirm the expert estimates of the ergonomics factors with the TAWS use.

The performed SLS tests showed that piloting by the FNI position bars during SLS-based approaches with the use of the DDRE and the OMS-Indicator is practically the same as for ILS-based approaches and does not cause difficulties for the pilots.

The integral evaluation of the flight safety improvement with the use of the methods and devices considered in this paper shows that the overall relative increase in flight safety determined in accordance with the indicators considered in Sect. 1.6, will be from 1.5 to 10 times.

Further development and improvement of the presented systems takes place in accordance with the principles stated in this section and today there are already modifications of the products considered above, which in many respects surpass those considered above.

References

1. Sauta OI (2011) Use of satellite navigation information to improve the efficiency of armament. Interuniversity Scientific and Technical Seminar. Abstracts of the report. Mikhailov Military Artillery Academy (in Russian)
2. Ageeva VM, Pavlova NV (1990) Aircraft instrumentation complexes and their design. In: Petrov VV (ed). Mechanical Engineering, Moscow, 432 p (in Russian)
3. US Patent No. 6 983 206B2, Cl. G08B 23/00, application of 15.05.2003, publ. on 03.01.2006
4. Babich OA (1981)Aviation instruments and navigation systems. Approved by the commander-in-chief of the Air Force as a textbook for students of Air Force engineering universities: Zhukovsky Air Force Engineering Academy, 348 p (in Russian)
5. Kalashnikov VS, Kuklev EA, Razumov AV, Olyanyuk PV, Sauta OI (2012) Principles of construction of onboard navigation and information systems (in Russian)
6. Kalashnikov VS, Kuklev EA, Olyanyuk PV, Plyasovskikh AP, Sauta OI (2012) Recommendations on the integrated use of satellite technologies in control systems for precision weapons and unmanned aerial vehicles (in Russian)
7. Sauta OI (1991) Use of SRNS beacons to support instrument approaches. In: Proceedings of the XI industry scientific and technical conference of the Ministry of Radioelectronic Industry, Leningrad, VNIIRA, pp 67–76 (in Russian)
8. Standards for Processing Aeronautical Data. RTCA DO-200A. Radio Technical Commission for Aeronautics, 2005
9. Operation manual. Differential data reception and conversion equipment. DDRE.NGTK.461534.001 RE. St.-Petersburg, VNIIRA-Navigator, 2006 (in Russian)
10. Baburov VI, Ponomarenko BV (2005) Principles of integrated onboard avionics. S.-Pb., Publ. House “Agency RDK-Print”, 448 p (in Russian)
11. Qualification requirements “Onboard equipment for satellite navigation” QR-34–01, 4th edn. Interstate Aviation Committee, 2011 (in Russian)
12. Baburov VI, Volchok JG, Galperin TB, Gubkin SV, Dolzhenkov NN, Zavalishin OI, Kupchinsky EB, Kushelman VY, Sauta OI, Sokolov AI, Jurchenko JS (2008) Airplane landing method using a satellite navigation system and a landing system based thereon. Patent No. RU 2 331 901, publ. 20.08.2008, Bul. No. 23 (in Russian)
13. Baburov VI, Volchok YuG, Galperin TB, Gubkin SV, Sauta OI, Sokolov AI, Jurchenko YuS (2010) Airplane landing method using a satellite navigation system and a landing system based thereon. Patent No. RU 2 385 469, publ. 27.03.2010, Bul. № 9 (in Russian)

14. Baburov VI, Volchok YuG, Galperin TB, Gubkin SV, Sauta OI, Sokolov AI, Chistyakova SS, Jurchenko YuS (2009) Airplane landing method using a satellite navigation system and a landing system based thereon// Patent No. RU 2 371 737, publ. 27.10.2009, Bul. № 30 (in Russian)
15. Baburov VI, Volchok YuG, Galperin TB, Gubkin SV, Ivantsevich NV, Sauta OI, Sokolov AI, Chistyakova SS, Jurchenko YuS (2010) Satellite radio navigation landing system using pseudo-satellites. Patent No. RU 2 439 617, publ. 10.01.2012, Bul. № 1 (in Russian)
16. Baburov VI, Galperin TB, Gerchikov AG, Ivantsevich NV, Sayuta OI, Sokolov AI, Chistyakova SS, Jurchenko YuS (2012) Complex method of aircraft navigation. Application No. 2012136399, priority of 17.08.2012 (in Russian)
17. Sauta OI (1986) Obtaining estimates of statistical characteristics of SRNS measurements on the approach trajectories. ChTP, State reg. No. I-85628 (in Russian)
18. Baburov VI, Volchok YuG, Galperin TB, Gubkin SV, Maslov AV, Sauta OI (2007) A Method for preventing an aircraft collision with a terrain relief and a device based thereon. Patent No. RU 2 301 456, publ. on 06.20.2007, Bul. No. 17 (in Russian)
19. Baburov VI, Volchok YuG, Galperin TB, Gubkin SV, Maslov AV, Sauta OI (2009) Method for preventing aircraft and helicopter collisions of with terrain and a device based thereon. Patent No. RU 2 376 645, publ. on 20.12.2009, Bul. No. 35 (in Russian)
20. Baburov VI, Volchok YuG, Galperin TB, Gubkin SV, Maslov AV, Sauta OI (2011) Method of notification of the aircraft location relative to the runways during the approach. Patent No. RU 2 410 753, publ. on 01/27/2011, Bul. No. 3 (in Russian)
21. Baburov VI, Volchok YuG, Galperin TB, Gubkin SV, Maslov AV, Sauta OI, Rogova AA (2011) Method of notification of the aircraft location relative to the runways during the approach and roll-on operation. Positive decision on granting a patent following the application for an invention No. 2011113706/11 (020353), priority of 04/04/2011 (in Russian)
22. Gomin SV, Makelnikov AA, Maslov AV, Priyemysheva AA, Sauta OI, Semenov GO, Sobolev SP (2005) Program for warning and emergency alarms generation. Certificate of official registration of computer programs. RF №2005611918. Federal Service for Intellectual Property, Patents and Trademarks, 2005 (in Russian)
23. WWW.JAVAD.COM
24. Sauta OI (1991) Substantiation of the coordinate system selection in the construction of algorithms for complex information processing in the landing system based on the short-range radio technical navigation system (in Russian)
25. Sokolov AI, Yurchenko YuS (2010) Radioautomatics: textbook for students of universities. Publ. Center "Academy", 272 p (in Russian)
26. AVIATION REGULATIONS (2004) Part 25. Airworthiness requirements for transport aircraft. In: Interstate Aviation Committee. OJSC "AVIAIZDAT", Moscow (in Russian)
27. AVIATION REGULATIONS (1995) Part 29. Airworthiness requirements for transport rotorcraft. In: Interstate Aviation Committee. OJSC "AVIAIZDAT", Moscow (in Russian)
28. Sauta OI, Gubkin SV (1987) Study of the mathematical model of the SRNS onboard receiver input signal in the landing area. In: Radioelectronics issues. Series OVR-1987-Issue 5, pp 39–48 (in Russian)
29. Annex 10 to the Convention on International Civil Aviation. Aeronautical Telecommunications. Radio Navigation Aids, 6th edn, vol 1, July 2006 (in Russian)
30. Act on the results of ground and flight tests of the Yak-42D main aircraft of airline "GAZPROMAVIA" equipped with onboard GNSS/LAAS equipment. Approved by the Federal transport inspection service of the Russian Federation. Moscow, 2006 (in Russian)
31. Certificate of the product conformance. SGKI-034-112-SRPBZ. Interstate Aviation Committee. Aviation register, 2003 (in Russian)

Correction to: Development of Navigation Technology for Flight Safety

Correction to:
Baburov S.V. et al., *Development of Navigation Technology for Flight Safety*, Springer Aerospace Technology, https://doi.org/10.1007/978-981-13-8375-5

In the original version of this book, the affiliation of the author "Bestugin A.R." was incorrect. This has now been corrected as follows:

Bestugin A.R.
Saint-Petersburg State University of Aerospace Instrumentation (SUAI),
St. Petersburg,
Russia

The updated version of the book can be found at
https://doi.org/10.1007/978-981-13-8375-5

© Springer Nature Singapore Pte Ltd. 2020

Baburov S.V. et al., *Development of Navigation Technology for Flight Safety*,
Springer Aerospace Technology, https://doi.org/10.1007/978-981-13-8375-5_7

Conclusion

This monograph describes an integrated approach to solving the issues of improving flight safety (FS) of manned and unmanned aerial vehicles (aircraft) at low altitudes with the help of technologies of global navigation satellite systems (GNSSs) used in collision avoidance systems (CASs) and satellite-based landing systems (SLSs).

The main results obtained on the basis of the presented approach can be briefly formulated as follows.

An approach is proposed for SLS and CAS construction that is distinctive in that the selection of the optimal alternative to be used as the functional core of these systems is based on a multiple factor analysis using the fuzzy sets theory. Such an approach makes it possible to trace the dynamics of the optimization process, to identify groups of factors that dominate the decision-making process, and also to increase the volume of influencing factors and to take the correlation relationships between individual parameters and criteria into account. The proposed approach allows consistent selection of key elements for radiotechnical complexes of various navigation systems, significantly shortens the time required to find the required solution, and, on the whole, accelerates the process of developing new equipment by 1.5–2 times. With the use of the proposed approach, new structures for constructing instrument SLSs and CASs have been developed, in which information from the GNSS and local ground-based augmentation systems is used. The proposed structures for the construction of radiotechnical complexes and systems, in contrast to the previously known ones, include software and hardware modules that enhance the accuracy, continuity, and integrity of navigation information. As a result, this ensures safety improvement for flights at low altitudes and over hazardous terrain during approaches and landings.

The theoretical foundations for the construction and use of the volumetric diagrams for distribution of errors in multipath propagation of radio waves are developed, which are refined on the basis of statistical accumulation of information on the characteristics of the received signals on the LAAS, which makes it possible to increase the accuracy of estimating the LAAS differential corrections by a factor

© Springer Nature Singapore Pte Ltd. 2020

Baburov S.V. et al., *Development of Navigation Technology for Flight Safety*, Springer Aerospace Technology, https://doi.org/10.1007/978-981-13-8375-5

of 1.5–6.0 times in the presence of re-reflected signals. At the same time, the false alarm probability is reduced by several orders of magnitude, the continuity of the landing signals output is increased, and as a result, the integrity of the data and flight safety is increased.

A method has been developed to improve the integrity and continuity of navigation information for the SLS, which is based on taking into account ground-based GNSS augmentations with mid-frequency variations of navigation parameters due to re-reflections, and the use of the predicted values of the signal-to-noise ratio determined in the GNSS receiver. The proposed method allows maintaining the established accuracy of generated coordinate information in the presence of unintentional electromagnetic interference during the landing, as well as during the landing of aircraft or UAVs in areas with a complex electromagnetic environment. The method makes it possible to increase the accuracy of determining the coordinates by 2 or more times. It should be noted that the errors in the determination of navigation parameters compensated by this method are not taken into account by standard algorithms of LAAS integrity monitoring.

A method for compensating errors in the determination of navigation parameters (pseudoranges) in the ground and onboard SLS subsystems has been developed on the basis of algorithms for detecting jumps in phase measurements using fail-safe filters. In comparison with known methods, the proposed method helps detect frequently occurring failures (cycle slips) in the operation of tracking systems that lead to jumps in navigation parameters. The new method for detecting jumps in phase measurements uses phase measurements with compensated noise of the GNSS receiver reference oscillator and helps detect multiple failures in a group of satellites. The application of the proposed method makes it possible to increase the accuracy of determining the aircraft coordinates by 3–4 times in the presence of various types of electromagnetic interference in the landing area for low-altitude flights.

A method has been developed for the application of a GNSS navigation receiver in the SLS using GNSS augmentation systems based on pseudosatellites (PSs). The proposed method is based on the reception of PS radio signals at a given (arbitrary) frequency and retransmitting it with automatic power limitation at the operating frequency of the GNSS receiver. The method resolves the main theoretical and practical problem that prevents the wide introduction of promising GNSS augmentation systems based on the PSs, that is, a large dynamic range of signal level variation that precludes the use of conventional GNSS receivers for simultaneous reception of signals from navigation satellites and PSs. The proposed method improves the accuracy and continuity of navigation information in the CAS by several times. When using one PS, the accuracy of navigation solutions can increase by 2–3 times, especially in conditions of a complex electromagnetic interference situation.

A method for preventing collisions of manned aircraft with ground for low-altitude flights has been developed based on geoinformation technologies, GNSS technologies, and a mathematical method for three-dimensional synthesis of the underlying surface sections with the identification of hazardous terrain elements or artificial obstacles. Expert estimates show that this method allows reduction of the time for making a decision on the nature of the necessary maneuver for safe flight continuation by 2.6 times and also helps increase the clearance margin over the terrain or obstacle by 1.6 times in the case of making a decision to recover from the hazardous situation by climbing.

A method has been developed for deciding whether to perform a maneuver by an aircraft pilot or a UAV automatic control system providing an integrated approach to selecting the safest flight path, taking into account constraints imposed by the nature of the terrain, the aircraft performance, and other available information. This method allows reduction of time required for the pilot to decide on the nature of the necessary maneuver by up to 3.5 times.

A method for preventing collisions with the ground has been developed based on the use of dynamic equations of aircraft or UAV movement and the analysis of the nature of the underlying surface in the flight area. The use of dynamic equations of movement makes it possible to construct mathematical models for calculating the minimum permissible turn radii and the predicted flight path, and the terrain scanning algorithms based on geoinformation databases and current parameters of the aircraft or UAV movement obtained from the GNSS help determine the degree of hazard when flying along the predicted path. The use of this method makes it possible for the pilot to reduce the time needed to make a decision on how to perform the necessary maneuver by 1.6 times. The method can be applied, among other things, for aircraft with the capability to take off and hover vertically, like helicopters.

A method for preventing aircraft landings on an unauthorized runway has been developed, based on the calculation of virtual landing glide paths. In the proposed method, in contrast to the previously known ones, generation of an approach runway selection notification is provided that is adapted to the type of aircraft, as well as generation of alerts/warnings of the aircraft deviations from the intended trajectory, and deviations of the predicted landing point from the intended one. Expert estimates show that the use of the proposed method allows reduction of the time required for the pilot to decide on the nature of the necessary maneuver to prevent an emergency situation by 4.2 times.

Complex methods for providing alerts/warnings of the aircraft or UAV position during the landing and the roll-on operation have been developed. These methods provide enhanced SLS capabilities through CAS functions and support for CAS functions through the use of SLS navigation information. Unlike the previously known methods, the newly developed methods use substantiated parameters (distance from the runway or target, glide path slope) of switching from angular deviations from the intended trajectory to linear deviations. In the new methods, alerts/warnings are generated of an unacceptable lateral deviation from the runway centerline during the roll-on operation; the distance remaining to the runway end is displayed, as well as the projections of the required and actual braking vectors, and the projection of the calculated aircraft stop point onto the runway centerline. The use of the developed complex methods provides substantial increase in the accuracy of navigation data used in the CAS and reduction in the false alarm probability by several times.

The principles of constructing modern systems for improving flight safety are justified. First of all, these are: multifunctionality; multiple systems; integration; multiple channels; parallel operation and redundancy; and testability and controllability. The design feature of the construction of SLS- and CAS-based flight safety enhancement systems is the use of the following common devices: antenna-feed system, multipurpose displays, and onboard GNSS receivers. The optimal alternatives that determine the functional core of any modern radioelectronic complexes should be selected using the methodological approach developed in this paper.

On the basis of the developed methodological approach in the process of development of new aircraft equipment samples, the influence of subjective factors on the selection of key functional elements for building flight safety systems is eliminated to a large extent, and the procedure for selecting these elements is simplified by using the new approach based on the use of expert navigation-oriented systems using a wide range of preference criteria and the mathematical apparatus of the fuzzy sets theory. The use of the proposed approach resulted in the equipment of Russian aircraft with domestic competitive equipment and the termination of the use of imported avionics regarding SLSs and CASs.

The developed methodological approach allowed reduction of the time frame for the development of new SLS and CAS models by 1.5–2 times through the use of the GNSS navigation field and its augmentation systems. The proposed structural diagrams for the SLS and CAS construction made it possible to modernize about 30 types of flight navigation instrumentation of aircraft in operation and thereby to improve flight safety.

The developed onboard SLS and CAS subsystems built on the basis of the proposed methodology (rules, techniques, ways, methods, algorithms) have now been implemented in many software and hardware packages of modern avionics products. It should be noted that in the last ten years, there have been no accidents with aircraft equipped with such products.

The use of the recommendations given in this paper makes it possible to reduce the false alarm probability in collision avoidance systems by 4–5 times when these are used together with satellite-based landing systems by increasing the accuracy of determining navigation parameters in the GNSS differential mode by 4 or more times, thus providing a reduction in the accident probability in most accident-prone flight phases, approach, and landing.

Application of the developed approaches in terms of increasing the ergonomics of the onboard navigation instruments by providing the crew with the image of the underlying surface together with the flight information can significantly improve the landing safety with the joint SLS and CAS use. Therefore, when creating new types of onboard systems for improving flight efficiency and safety, these two systems should always be operated on the basis of integrated use.

The materials of the monograph can be used to develop technical specifications for onboard flight navigation systems (complexes), for ground GNSS augmentation systems, and also for the development of navigation radioelectronic complexes to enhance their efficiency.

Most of the methods and systems presented in the monograph are protected by RF patents and international applications for inventions.

The developed methods and systems, as of the beginning of 2016, were used in onboard radioelectronic equipment of more than 2000 aircraft of the state and civil aviation of Russia.

Baburov Sergey Vladimirovich, Candidate of Technical Sciences
Bestugin Alexander Roaldovich, Doctor of Technical Sciences
Galyamov Andrey Mikhailovich, Candidate of Technical Sciences
Sauta Oleg Ivanovich, Doctor of Technical Sciences
Shatrakov Yurii Grigorievich, Doctor of Technical Sciences, professor, honored science worker of Russia, laureate of state and St. Petersburg government prizes, RATS full member

The paper addresses issues of improving the safety of low-level flights. It considers the perspective scientific direction on the creation of integrated systems for supporting navigation and landing of aircraft performing low-level flights, including takeoff and landing operations. The authors focus on promising technologies for creating integrated safety systems based on various navigation fields. The monograph is intended for specialists in radioengineering: radiolocation, radionavigation, navigation and air traffic control, system analysis, development and modeling of radiotechnical systems.

MIX
Papier aus verantwortungsvollen Quellen
Paper from responsible sources
FSC® C105338

If you have any concerns about our products,
you can contact us on
ProductSafety@springernature.com

In case Publisher is established outside the EU,
the EU authorized representative is:
Springer Nature Customer Service Center GmbH
Europaplatz 3, 69115 Heidelberg, Germany

Printed by Libri Plureos GmbH
in Hamburg, Germany